T0073857

# What Is Regeneration?

## CONVENING SCIENCE

DISCOVERY AT THE MARINE BIOLOGICAL LABORATORY

**A Series Edited by Jane Maienschein**

For well over a century, the Marine Biological Laboratory (MBL) has been a nexus of scientific discovery, a site where scientists and students from around the world have convened to innovate, guide, and shape our understanding of biology and its evolutionary and ecological dynamics. As work at the MBL continuously radiates over vast temporal and spatial scales, the very practice of science has also been shaped by the MBL community, which continues to have a transformative impact the world over. This series highlights the ongoing role MBL plays in the creation and dissemination of science, in its broader historical context as well as for current practice and future potential. Books in the series will be broadly conceived and defined, but each will be anchored to MBL, originating in workshops and conferences, inspired by MBL collections and archives, or influenced by conversations and creativity that MBL fosters in every scientist or student who convenes at the Woods Hole campus.

# What Is Regeneration?

Jane Maienschein
and Kate MacCord

The University of Chicago Press

Chicago and London

The University of Chicago Press, Chicago 60637
The University of Chicago Press, Ltd., London
© 2022 by The University of Chicago
All rights reserved. No part of this book may be used or reproduced in
any manner whatsoever without written permission, except in the case
of brief quotations in critical articles and reviews. For more information,
contact the University of Chicago Press, 1427 E. 60th St., Chicago, IL
60637.
Published 2022
Printed in the United States of America

31  30  29  28  27  26  25  24  23  22      1  2  3  4  5

ISBN-13: 978-0-226-81656-2  (paper)
ISBN-13: 978-0-226-81657-9  (e-book)
DOI: https://doi.org/10.7208/chicago/9780226816579.001.0001

Library of Congress Cataloging-in-Publication Data

Names: Maienschein, Jane, author. | MacCord, Kate, author.
Title: What is regeneration? / Jane Maienschein and Kate MacCord.
Other titles: Convening science.
Description: Chicago ; London : The University of Chicago Press, 2022. |
    Series: Convening science: discovery at the Marine Biological
    Laboratory | Includes bibliographical references and index.
Identifiers: LCCN 2021035851 | ISBN 9780226816562 (paperback) |
    ISBN 9780226816579 (ebook)
Subjects: LCSH: Regeneration (Biology)
Classification: LCC QH499 .M26 2022 | DDC 571.8/89—dc23
LC record available at https://lccn.loc.gov/2021035851

♾ This paper meets the requirements of ANSI/NISO Z39.48-1992
(Permanence of Paper).

# Contents

# Preface

In 2016, President Susan Fitzpatrick of the James S. McDonnell Foundation challenged us to think about how to "put history and philosophy of science to work with the life sciences." We had been arguing that historical perspective and philosophical analysis (or HPS studies) can improve the life sciences. Susan asked us to show how. That led to a workshop, which led to a foundation grant to explore different ways we could carry out this multidisciplinary work. We set up working groups for biodiversity, cell biology, and developmental biology, and all three led to larger projects. Yet each remained focused on particular areas of biology. Susan urged us to find a topic that spans all of biology, asking what themes hold for all scales of life. We agreed that we would focus on the community at the Marine Biological Laboratory (MBL) in Woods Hole, Massachusetts, where both of us have roles.

It turns out that biologists can become both excited by the prospect of looking at life across different scales and worried that the task is too big and too bold. As a result of many conversations and working group interactions, we chose the topic of regeneration. For developmental and cell biologists, regeneration of organisms and parts is a familiar topic, but they didn't all agree about the limits of what should count as regeneration: is it regeneration when an embryo recovers from damage? For neurobiologists, regeneration connects with hopes for regenerative medicine, but it is less clear how much can regenerate and whether regeneration of some cells can fix a damaged nervous system. Ecosystems ecologists expressed uncertainty: does dune restoration count as regeneration? Microbiologists asked whether restoring a microbial community means regeneration or some kind of replacement? It became clear that biologists have more questions than answers about regeneration.

This is where HPS perspectives can contribute much to the research. We decided to bring together five working groups, each including at least one historian, philosopher, and biologist. What does regeneration mean and how has it been studied in, for example, neurobiology, stem cell biology, germ line cell studies, ecosystems ecology, microbial evolution? And what do we learn by looking at all of these together? What is regeneration and, especially, what factors drive and what rules govern regenerative processes? Is there one logic of regeneration, or set of rules, across all of these different areas of biology and different scales of life?

Opening these questions to greater exploration in different areas has been a lot of fun. Working together, meeting regularly to compare ideas and probe suggested interpretations has led to new insights. We're bringing those together here.

Jane Maienschein is a historian and philosopher of the life sciences, especially of developmental and cell biology, who looks closely at the scientific details while also asking about the broader social context. She directs the Center for Biology and Society at Arizona State University and the History and Philosophy of Science program at the MBL. She directs the James S. McDonnell Foundation grants. Kate MacCord is a historian and philosopher of biology who received her PhD from Arizona State and serves as codirector for the McDonnell projects. She coordinates the working groups and thus helps to promote lively discussion that digs into the hard questions. Together, we raise questions, offer suggestions, and open discussion that will continue through the books that will follow in this series of short books, in workshops, and in the halls and labs at the MBL and elsewhere. We invite you to join the fun and help refine the questions and offer additional suggestions.

# 1     The Idea of Regeneration

Picture yourself getting up early, eagerly looking forward to a bagel and coffee. Unfortunately, you cut your fingers slicing the bagel, then burn your hand making the coffee. Most likely your skin will heal, undergoing regeneration to replace the damaged cells. Soon, you will have forgotten all about the cut and burn. Or maybe you will get an infection and have to take strong antibiotics. These antibiotics upset your gastrointestinal system by killing off the microbes that normally help you digest your food and turn it into useful energy. Luckily, in most cases, something like yogurt or other "probiotics" will repair the system so that it regenerates its function to get your digestion working normally again. Then, to make yourself feel better, you might go to your favorite forest campground. A big fire killed off many trees and has left a damaged landscape, but new seedlings are already springing up in signs of renewal and regeneration of life.

In all of these cases, an injured system—like your fingers, your digestion, and the forest—undergoes an adaptive response that restores the structure and function of the whole system by repairing the parts. In what ways are these types of regeneration the same or different? How can we learn from regeneration within each of these living systems to compare and translate the knowledge to other kinds of systems? Policymakers, biologists, historians, philosophers, students, teachers, and general readers will all benefit from understanding regeneration and envisioning how the process carries across all scales of life. The intention of this book is to help readers gain a new perspective on regeneration as a process of all living systems. This new perspective will also aid recognition of how all living systems are interconnected and impact each other.

By recognizing similarities and connections that allow us to apply knowledge of regeneration from one scale of living systems to others, we may be able to treat debilitating degenerative diseases and injuries and even heal our fractured planet. This book offers a short introduction to the concept of regeneration across living systems, with the intention of introducing the idea to all readers who care about learning ways to repair damage in the future and who understand that the future is informed by the present and the past. While this book focuses on regeneration in individual organisms, it is the first of a series that will provide further examination at different scales of life including stem cells, germ line cells, nervous systems, microbial communities, and ecosystems.

In this book, we advocate for a "systems-based" approach to understanding regeneration. What, then, do we mean by "system" in this context? A system is a group of parts that interact in a coordinated fashion. The resulting whole follows rules and principles, which allow some kind of communication and integration of the parts so that the entire system is responsive and regulated. Therefore a "systems-based" approach seeks to understand the parts of systems, their interactions, and the rules that govern them. Systems can be defined at nearly any scale. The microbes in your gut make up one system, the cells that coordinate to heal your finger make up a different system, and all of the species and inorganic materials within your favorite campground make up yet more systems.

Our approach might be confused with the field known as "systems biology"; however, while systems biology uses computational and mathematical analyses to model complex biological systems, we are up to something quite different. We are aware that biologists who study different scales of systems often do not even think about how their findings relate to those at other scales: ecosystem ecologists are sometimes puzzled when we suggest that their forest regenerates in ways that are like the regeneration of skin in the finger, for example. Yet by looking at the regulated and integrated whole that results from the parts, we see many similarities and therefore a value in exploring comparisons across scales of life. To do that, we want to push our readers (and ourselves) to reflect on how we have come to think about regeneration in certain ways and in different systems (through history), and how we know what we know about regeneration (through philosophy).

Studies of regeneration have given us a huge amount of knowledge about different aspects of regeneration within living systems, especially in individual organisms. They have, however, also remained reductionistic in looking at parts and details rather than whole systems. For example, many studies look at particular nerve cells and the ways they work but not at the whole nervous system. Others look at how particular species regrow in a forest after a fire, for example, but not at the entire ecosystem and the way a large interacting community of different species interacts together and within the changing environment. In this case, focusing narrowly makes it much less likely that we can even see the whole forest, much less get a sense of how it will affect the future and what we need in order to respond to climate change and other forms of system damage.

In order to compare and begin to learn from our growing knowledge of regeneration within different living systems, we need first to determine whether there are sets of rules or what we call a "logic" that governs regeneration at each particular scale. All systems, regardless of scale, have generalizable characteristics—types of parts within the systems, types of relationships between the parts, types of relationships with the environment, and sets of rules that govern their interactions. Types of parts could be groups of cells or molecules within a regenerating limb or particular species within a microbial community or an ecosystem. The roles or functions they play within the regenerating system define what we call their "types." Within an organism, the parts interact with each other in interactive relationships, and as a result some cells within the regenerating limb may initiate regeneration, activating others to proliferate, while others regulate how those proliferating cells form into replacement tissues. Within a microbial community, the different species of microbes interact in similarly complex ways, with the loss of some species leading to replacement by others.

Because interactions and parts can be understood as "types," they follow patterns and sets of rules acting as a logic that dictates how the parts and relationships work during regeneration of lost or damaged structures or functions. Because regeneration takes place within a system, and all systems have such generalizable characteristics, *every* living system has a logic of regeneration that governs how and under what conditions regeneration proceeds. This logic may not be identical for *all* living systems, though that remains an open question that this series of books is exploring. The sets of rules, or logics, nonetheless surely have features in common and we initiate and expand questions about the nature of those shared features. We ask whether regeneration means the same thing at different scales: are all of the different types of regenerating systems doing the same thing or something different? Is there an underlying logic of regeneration across all of these different types of living systems or different logics?

Some traditional biologists will surely ask "so what?" Given that life sciences have been successfully making many advances through the typical reductionist approach that focuses on different scales in isolation, why bother to take a larger systems-based approach in which we search for and compare the logics of regeneration across living systems? Because, we believe, comparing the logic by which regeneration operates *within* and *across* living systems as different as microbial communities, neurons, and forests opens up a realm of possibilities for making novel and unexpected discoveries about each system as well as about life in general.

For example, one aim of regenerative medicine is to restore function to severed spinal neurons in humans that lead to paralysis. Scientists have already gained many insights by looking at spinal neuron regeneration throughout a diverse set of organisms—from mice to lampreys—and comparing how regeneration works in each. This perspective from the study of diversity is leading scientists to understand what is missing in humans that keeps us from regenerating spinal neurons. Part of the answer about what is in common lies with genes and gene expression, and part lies with the ways that the systems' parts interact.

Now imagine if we expand our perspectives even further and compare spinal neuron regeneration to how the gut microbiome restores healthy human function by creating new community structures with the same function as the preregenerated community. If we can abstract the logic of regeneration in one system and compare how it operates within each other type of system, we should be able to gain even more insights into what we're missing from our understanding of spinal neuron regeneration, and maybe even ideas about how to reengineer our system in order to regenerate more effectively.

Another benefit of thinking across systems at different scales comes back to the connectedness of all living things. Organisms and ecosystems act as hosts for microbiomes, which in turn provide functions critical to the health and well-being of the microbes, organisms, and ecosystems. Ecosystems comprise vast networks of organisms, which interact and affect the lives of each other. Life on our planet is a network of living systems that all affect each other, which we see most clearly in the global effects of humans during the current era that is called the Anthropocene because it is so dominated by human actions as a global force. Because all living systems are connected, it would be very shortsighted to think we could fully understand regeneration within one system without understanding it at other scales.

By committing ourselves to thinking of regeneration as a phenomenon that is at least shared in some respects by all different types of living systems and as being governed by a shared set of rules or logic, we are committing ourselves to the goal of making knowledge translatable and transferrable within and across living systems. Transferrable knowledge is an ambitious but necessary goal if we want to help regenerative medicine achieve its aims and if we want to understand our planet enough to begin counteracting damages of the Anthropocene. This short book does not offer answers or provide a grand unified theory of regeneration. Instead, it expands questions and challenges us to think about what regeneration is across all scales of life.

Regeneration is not a new concept, though it started with a focus on individual organisms and their parts. Greek mythology's Prometheus saw firsthand the curse and blessing of being able to regenerate organs. Always creative when doling out justice, the gods punished Prometheus for stealing fire and giving it to mortals by chaining him to a rock, where he had to ward off a daily eagle attack. Every day the eagle plucked out his liver and every night the liver regenerated, so Prometheus could live to suffer again. The human liver does, in fact, have the capacity to regenerate its cells and thereby to recover its function. Surely the ancient Greeks didn't know this, but the truth in their story is intriguing. Some read the story of Prometheus with horror at the fate of constant injury, while others embrace the hopefulness of constant repair. For many, the idea of a modern Prometheus in which regenerative medicine allows us to regrow and repair any bodily damage invokes ambivalence; playing with fire, after all, can be useful or dangerous (fig. 1.1).

The ancient Greeks did not relegate regeneration solely to mythology nor did later observers. They also witnessed and documented regeneration within nature. In his extensive study of natural history during the fourth century BC, the philosopher Aristotle investigated regeneration in snake and lizard tails in particular. Aristotle was a careful and creative observer, full of curiosity about the world. He noticed how a snake or lizard that lost its tail could grow a new one—and he developed the first known theories about what causes regeneration in some cases but not others, drawing on his understanding of the world as guided by a set of four types of causes. Other ancient observers asked similar questions, noting that hydra and earthworms also regenerate.

Over time, and with more and more people adopting the naturalist's interest in observing nature, it has become clear that some organs, tissues, and even organisms can regenerate parts or even their entire living systems under some conditions, while others cannot regenerate under any conditions. Cut an earthworm in half in just the right way, and two wiggly worms may confront you. Cut off the head of a hydra, and it

FIGURE 1.1 | "Regenerative Medicine," illustration by Terese Winslow. Featured as the cover of the 2006 NIH report on regenerative medicine.

may grow a new head. Cut off a human arm, and the human will always be missing an arm. At another scale, a damaged microbial community can sometimes undergo restoration of its function. Forests are sometimes, but not always, said to undergo restoration or regeneration after a fire. And in the face of climate change in the Anthropocene, we might go to an even larger scale and wonder whether the planet can regenerate to become healthy again or whether the global ecosystem will collapse.

Each of these examples of systems is viewed as an independent case where regeneration may be possible and yet, again, we can also see that there are connections among these systems. Microbiomes, for instance, reside within earthworms, hydra, humans, and forests. Humans are the main agents driving many changes during the Anthropocene and affecting the health of the global ecosystem. The connectedness of living systems means that however we carve nature into independent units and seek to understand systems at different scales, they will always have an impact on each other. Seeing some form of regeneration in these different scales of living systems therefore raises questions about the causes, limits, connections, and possibilities.

And yet, when we look at the way different researchers have seen regeneration, we find that they have not always used "regeneration" to mean the same thing. Sometimes regeneration means bringing a system back after damage or injury, as with re-generation, re-juvenation, re-vitalization, re-newal, re-mediation, and re-silience. Sometimes the definition emphasizes the result or final state, such as with re-pair, re-storation, re-plication, re-covery, re-placement, and re-silience. Researchers have also had different ideas about what is capable of regenerating. Apparently, it is living systems that regenerate, but then we have to know what counts as a living system. And if it is parts of systems that regenerate, then there are different ideas about what count as the parts of a system. Different ideas also hold for what counts as a normal state against which to assess injury or damage, as well as how to recognize whether regeneration has led to restoration and repair.

To get at a range of different understandings of how regeneration plays out in different living systems, we look here at how the meaning of regeneration has shifted over time. We introduce well-known eighteenth-century biologists, such as René-Antoine Ferchault de Réaumur and Abraham Trembley, who empirically studied regenerating organisms, such as a genus of small, freshwater organisms called *Hydra*. They were largely exploring, observing, describing, and beginning to wonder how different organisms could regenerate some parts but not others. By looking historically at these leading examples, we also introduce philosophical questions about how these studies should be done, what counts as evidence, and even whether one can determine what is normal by cutting off heads and other parts.

Then we show how this early work evolved with research by twentieth-century biologists Thomas Hunt Morgan, Jacques Loeb, and Charles Manning Child, looking at the ways that they embraced experimentation to study regeneration, how they searched for mechanisms, and how they began to envision organisms as living systems. Their emphasis on regenerating organisms as biological systems emerged starting in the early twentieth century. Our empirical discussion of these historical ideas establishes the traditional context for thinking about regeneration, largely within individual organisms and their parts.

Building on historical studies of regeneration in individual organisms carries us to the latter half of the twentieth century and into the twenty-first century with an overview of increasingly diverse modern investigations of regeneration in different types of living systems. We begin with cells, specifically neurons, stem cells, and germ line cells. Each cell is, in and of itself, an individual living system. It is a discrete biological unit that can be separated from its neighbors within a tissue, isolated within a petri dish, and monitored under the microscope. It is also a mass of proteins and structures that communicate and interact in order to sustain its own necessary functions and allow it to operate with its neighbors to perform the functions necessary to sustain the life of the organism. The neuron, the stem cell, and the germ line cell are thus all discrete, living systems that are also units or parts of other, larger systems like organs, tissues, and organisms.

Individual cells come together to make up organisms as a different type of living system. Organisms, like cells, exist in larger environments and are parts of other, interconnected systems, including populations, communities, and ecosystems. Ecosystems contain everything from microbial communities that function as systems in their own right, to animals, plants, and inorganic items such as rocks. Each of these plays a role in sustaining the ecosystem's life. Systems and parts can therefore be defined at a variety of scales, from the microscopic to the organismal to the ecological, and how we define systems and their parts affects our understanding of regeneration within systems.

Determining that regeneration has occurred within a system

is usually a matter of comparing the states of a system before and after an injury. This is not a straightforward task. Let's go back to our first paragraph. You have a microbiome, a community of bacteria that helps you digest food—the initial state. You take antibiotics, which kill your gut microbiome—the injury. You then eat yogurt which returns bacteria to your gut and you can again digest foods—the final state. Is this regeneration? Before the antibiotics, your gut bacteria existed in a particular community structure, with particular sets of species interacting with each other in order to sustain your gut health. The result afterward may restore your system's digestive function and make you feel healthy again, but the microbial community is likely to be quite different. In this case, the system's function regenerates even though the details of structure are not restored exactly.

Regeneration in microbial communities or ecosystems is complex. It is just as complex in cells or organisms. For example, when the spinal neuron of a lamprey (a species of jawless fish) is severed, it will regenerate and restore the swimming ability of this basal vertebrate. Yet the structure of the neuron is not the same after regeneration as before the injury.[1] In another example, axolotls (an adorable species of salamander very popular with biologists) can regrow entire limbs and regenerate their skin without any scarring. Axolotls can regenerate their limbs and skin with apparently perfect structural and functional fidelity to their uninjured parts, while most systems cannot. In this case, the structure is the same after regeneration as it was before the injury occurred. Thus, the ways in which we think about the states of the system before and after injury, and whether we think of regeneration as restoration of structure and/or restoration of function, impact what we think regeneration is and whether and where it occurs.

Recall our core questions: are all these regenerative phenomena the same thing or different? And is there an underlying logic of regeneration across all these different types of living systems? As we dig into details, following the empirical and then interpretive discussions of regenerating living systems as they unfold, it is useful to think a bit about why it matters to understand regeneration. The systems we focus on throughout this book series—germ line cells, stem cells, neurons, microbial communities, and ecosystems—are the sources of a great deal of hope for repairing ourselves, for ensuring the safety and security of future generations, and for restoring our fractured planet, *if* we can get regeneration to work.

Think about germ line cells. Germ line cells are the reproductive cells (ova, sperm) that come from what is called the germ line and allow sexually reproducing species, such as ourselves, to have offspring. Since the late nineteenth century, biologists have assumed that these germ line cells are sequestered from the somatic, or body, cells and that they cannot regenerate. If they are damaged or changed in some way, the body cannot create new germ line cells and therefore cannot pass any damaged cells on to future generations. Hundreds of millions of dollars are spent annually in the United States on trying to understand why human germ line cells fail to regenerate and how this impacts the condition of infertility that affects many people. Germ line cells tend to be easily lost during the complex reproductive process or damaged by cancer therapies and other medical treatments, which contributes to making many people infertile. Yet biologists and medical researchers take it as an absolute given that germ line cells are a special fixed kind of cell that is separate and distinct from all other cells in the body and that once these germ line cells are lost, the body cannot transform somatic cells into germ line cells to regenerate them. As a result of these assumptions, many scientists have concluded that it is ethically acceptable to edit the genes of somatic (or body) cells but not these special germ line cells because they see no chance of edits introduced into the somatic cells becoming a part of germ line cells and being passed on to future generations.[2]

In this context, it is important to note that the assumptions about germ line cells are not true. Recent studies that draw on broad evolutionary comparisons suggest that some species actually have broad regenerative capacities in their germ line cells and that this is sometimes accomplished by turning somatic cells into germ line cells. The regenerative capacity of germ line cells has far-reaching implications when we consider the extent to which somatic cell genome editing is permissible in humans, but germ line cell genome editing is not, especially when we take into account that the extent to which, and the conditions under which, germ line cells can be regenerated from somatic cells in humans is unknown.

The implications of our lack of knowledge about somatic cells' ability to become germ line cells, when coupled with genome editing technologies, are staggering. For example, in laboratories around the world, scientists are laboring to edit genes involved in cancers. One such gene (*PD-1*) produces a protein that keeps the body from killing off cells. When this gene mutates, it can cause the immune system to malfunction and keep it from killing cancer cells. As of October 2020, three registered clinical trials in the United States were testing the efficacy of treating different cancers by using genome editing to knock out the *PD-1* gene in somatic cells. While *PD-1* is implicated in some cancers, it also plays a major role in normal immune system responses. If cells with the *PD-1* gene knocked out were to become germ line cells through a process like germ line cell regeneration, any children that received this genome edited cell would likely be severely immunocompromised. Therefore, better understanding the biology of how germ line cells regenerate will help us to clarify the ethical debates surrounding genome editing and allow us to make more informed decisions about reshaping our genomes and the future of our species.

We hope this book, and the set of rules or logic for identifying regeneration it outlines, will provide a foothold for understanding this and other related debates. It will also help readers understand scientific and ethical questions related to stem cells, which arise at the earliest cell divisions with some continuing to much later developmental stages. Stem cells play various roles at those later stages, and research over the past few years suggests that stem cells may be important for regeneration in individual organisms. Understanding their abilities to promote regeneration of structure and function can inform the possibilities for using them in clinical settings, where addressing medical problems is an enormously attractive — and lucrative — prospect. Only in 1998, with isolation of human embryonic stem cells did this idea of regenerative medicine begin to seem realistic. Since 1998, research on the use of stem cells to facilitate regeneration has skyrocketed, and institutes throughout the world have been formed that use stems cells in order to develop medical therapies.

After decades of research, researchers' lack of understanding about how this type of cell acts and its roles within regeneration impedes our ability to harness its power fully for medical purposes.[3] Despite this lack, the prospects for regenerating parts of the body, such as nerve cells, and restoring lost functions hold tremendous promise for many. Could we perhaps cure people with Parkinson's disease, with spinal cord injuries, with so many neurodegenerative disorders from which so many people suffer? Scientists have been working toward spinal cord regeneration since the turn of the twentieth century. With the advent of extensive advances in stem cell research in medicine and a broad perspective of spinal cord regeneration across all animals, we stand on the cusp of making this promise a reality.

Beyond individual organisms and their parts, regeneration in ecosystems matters even more in our era of the Anthropocene, wherein humans are clearly changing the climate and threatening living systems through fires, floods, food, air, water, energy, and other environmental impacts. Early in the twentieth century, American plant ecologist Frederick Clements theorized that plant communities might be analogous to organisms. He suggested that trees, other plants, and their nutrients might be similar to animals with their parts that work together as a living system. Clements's idea was supported in the 1950s by Eugene and Howard Odum, who became pioneers of eco-systems ecology. Arthur Tansley introduced the concept of ecosystems, to which we return later.

In this context, and especially in thinking about forest man-agement, researchers accepted that ecosystems are living sys-tems. They undergo damage, as after a fire, and then repair. Ecosystems are incredibly complex systems made up of mil-lions of interacting parts, and their health and regenerative abil-ities are affected by the systems with which they interact, such as the humans around and in them. They are parts of larger biosocial systems that make up our planet. So, how does this whole system of interacting parts regenerate something that actually works? If we want to combat the deleterious effects that humans have had on ecosystems, we need to understand the logic by which they can repair and regenerate. Only then can we begin to make informed decisions about how to inter-cede to help them regenerate their lost parts and functions.

As described in our discussions of the microbiome, microbes have in recent decades had an identity makeover to become more than just "germs" that energetic houseclean-ers should always work hard to eradicate. Microbes are often our friends. Indeed, communities of microbes make up over-lapping and interconnected microbial communities within our bodies that are in large part responsible for maintaining our health, as we saw with the example of gut microbes maintain-ing gastrointestinal health. Restructuring these communities or infiltration by species or variants of microbes can change the functions that they perform, leading to illness. We are only beginning to understand all the roles that friendly microbial communities play for us as individual humans. Different microbial communities work with other living systems as well. Agriculture relies on microbial communities to help make the soil fertile for the crops we want. Forests require microbial communities to maintain healthy ecosystems. Microbial communities change, they can be damaged, and they can recover.

By now, you have begun to grasp the extent to which regeneration occurs throughout living systems and why it is so important to understand what it means and how it works. Better understanding regeneration within and across living systems means moving away from our received understanding of how it is *supposed* to work. The systems-based approach we advocate throughout this book and series is meant to do exactly this: turn the complexity of the biological world into something we can interpret and then act upon and thus empower all of us—from the general public to the scientists performing experiments to the policymakers tasked with regulating science and our environment—to move toward restoring our health, making informed decisions about altering the future of our species, and mending a world fractured by the effects of the Anthropocene.

# 2     Observations and Experiments

Today the idea of regenerating living systems extends across the scales of life from individual cells and their parts to extensive ecosystems, but the earliest historical ideas concerned individual organisms. Organisms are visible, identifiable, apparently separated by boundaries such as skin from their surroundings, and capable of being injured. Empirical observations provided the starting point for early thinking about regeneration, with researchers exploring: seeing, documenting, and then comparing what appeared to be normal, what appeared to be abnormal due to damage, and how organisms could return from their abnormal or damaged state to a normal state. Since regeneration involves damage or injury to the normal, followed by recovery, these empirical observations provided a starting point for the idea that an individual living organism could be damaged and undergo repair through regeneration.

As occurs quite often in the history of scientific discovery, careful observations started in the fourth century BC with the Greek philosopher Aristotle. Also as per usual, Aristotle's accounts remained the standard until the modern period of scientific discovery in the seventeenth and eighteenth centuries identified with the Scientific Revolution and the Enlightenment. At that point, and especially in the eighteenth century, scientific investigators began to pay close attention to natural phenomena, especially the individual systems that are organisms, and to rethink some of the wisdom they had received from their predecessors, including Aristotle. In connection with their study of development from embryonic to adult form in individual organisms, researchers in the eighteenth century such as René-Antoine Ferchault de Réaumur and Abraham Trembley observed, recorded, and reported findings that showed the additional value of looking more closely and more extensively at both processes and products of the intriguing process of regeneration. The interest they helped spark among many other naturalists then led to a century and a half of discovery that Thomas Hunt Morgan summarized in his 1901 *Regeneration*.[1] There, as a retrospective report, Morgan brought together key contributions to the enthusiastic period of exploration and observation up to 1901 while also introducing his own extensive experiments and interpretations of regeneration. This chapter documents some of the history of these observations prior to Morgan's summary and explores how investigators thought about regeneration over the time period between Aristotle and Morgan.

## ARISTOTLE

Aristotle provided the background for later thinking with his theoretical accounts of empirical observations. That his interpretations gave way to other ideas after almost one and one-half millennia does not detract from the significance of Aristotle's initial contributions, so we begin with him. While he did not discuss regeneration at length or in detail, he did record observations about what we today think of as natural systems, and he placed them in the larger context of his highly organized interpretive framework.

Aristotle's impressively large corpus includes diverse studies that span the natural world, ranging from basic principles of physics and motion on Earth to the motion of the skies. For Aristotle, the world is not stable and fixed, though the underlying rules governing it are. Rather, the world results from the action of causes in ways that allow constant change or what he called generation. He had only to look up at the night sky to see that the stars move and to wonder why they move in such predictable ways. Aristotle's orderly mind led him to build on ideas of his predecessors and identify a world in which every object is made up of four elements: earth, fire, air, and water. The elements have qualities, so that each object is a combination of hot, cold, wet, and dry. And each is governed by four causes: the material, formal, efficient, and final.

While we point the interested reader to other resources for more detailed discussions of Aristotle's worldview, the important thing to keep in mind here is that Aristotle envisioned a world undergoing constant change and recognized that the regular causes that govern the movement of the heavens also govern the movements and changes of all life on Earth.[2] With his orderly worldview, Aristotle could provide an explanation for almost anything, including the development and regeneration of living organisms and their behaviors. Even though we no longer accept most of his explanations, we can still admire his observations and questions within his larger systematic interpretive framework.

Aristotle recognized that generation in animals, or the movement from unformed material to organized life-forms, included the process of restoration or regeneration in response to injury. For living organisms, Aristotle provided focused recordings of his observations and interpretations. As historian of philosophy James Lennox explains in "Aristotle's Biology," Aristotle clearly intended to apply the logical reasoning he had developed for physics to the living world. His *Parts of Animals* laid out the reasoning and appropriate approach for studying life, focusing on applying his four causes across different forms of living organisms. *Generation of Animals* looked at how animals originate and develop from conception to fully formed organisms. *History of Animals* provides rich descriptions of what Aristotle saw by looking at a wide variety of animals and includes reports of observations by others as well. Though Lennox explains that scholars are not exactly sure when these writings appeared, nor whether they were intended as polished products or more nearly lecture notes to record Aristotle's thinking, they offer amazingly detailed accounts of many aspects of living systems.[3] As many have said, Aristotle was a keen observer in a way that reminds us today to look carefully and record accurately what we see.

In *History of Animals*, Aristotle notes that "Some persons say that in one respect serpents resemble the young of the swallow, for if their eyes are pierced with a pointed instrument, they will grow again, and if the tails of serpents or lizards be cut off, they will be reproduced."[4] Here, Aristotle recognizes that the causes within the natural world must have the capacity to respond to injury that organisms experience. Once an individual organism has reached its fully functioning form, it does not completely stop changing. Rather, at least some organisms under at least some circumstances must have the ability to regenerate missing parts. He does not give an account of what this means in terms of his four causes, but given his discussions of generation more generally, we would understand this as a recognition of the motion and change within all parts of the living world. Responding to injury or the abnormal would be part of normal processes. Aristotle established the phenomenon and raised the questions about what occurs as well as how and why it occurs.

Aristotle's ideas found traction within the Roman Catholic church and provided the foundations for knowledge throughout western Europe for the period known as the Middle Ages. Some discussions occurred within medicine and concerned the aging process in humans, including hints of interest in what might restore youthfulness through some process of regeneration after loss or injury.[5] But the phenomenon of regeneration Aristotle had reported was not taken up again as a subject for serious observation and interpretation until the eighteenth-century period known as the Enlightenment.

## EIGHTEENTH-CENTURY NATURALISM

Many excellent histories of science have demonstrated the rise of what we would recognize today as modern scientific thinking, which entailed detailed observations, experimentation to extend the kinds of observations available, and thoughtful interpretations in the form of theories to be tested against additional observations. In the life sciences, such studies have looked at the tradition of natural history, which typically begins with observations across diverse organisms and asks many different kinds of questions about life. Naturalists explored the world, collecting objects and recording observations of the diverse things they saw. They brought many specimens back from voyages of discovery around the world, then built museums in which to house and to study those formerly living organisms. In order to make sense of what they had accumulated, naturalists began to ask what seem to us like the most basic questions, such as: What were all those different things? How were they related? How did each one work, and what made it normal or damaged in some way?

In the sixteenth century, curiosity drove early naturalists such as the German Conrad Gesner to record their amazement at discoveries from distant places. Gesner's astonishing volumes of natural history included folktales, fantastic descriptions of various monsters and curiosities, and also rich detail of many birds, frogs, mammals, salamanders, and others. Such a remarkable diversity of life-forms exists "out there," Gesner realized, and there was a sense that explorers were only beginning to appreciate all the amazing diversity. Gesner's volumes served at the time, as well as today, to cause us to marvel at the richness of the somewhat familiar as well as to wonder at the unique and perhaps fanciful.[6]

Two centuries later, the focus had become less about wonder and more about careful, recorded observations and interpretations of the phenomena of life. By then, biologists typically used microscopes to allow them to see more and to see it in greater detail. The renowned French naturalists Georges-Louis Leclerc, Comte de Buffon; Jean-Baptiste Lamarck; and Georges Cuvier produced dozens of volumes detailing the vast diversity and complexity of the natural world. Their catalogs of accumulating abundance of discoveries led to the formulation of many broad, encompassing theories that addressed the basic questions of their predecessors in an attempt to make sense of it all.

While some European naturalists marveled at the diversity of the world, fascinated by newly discovered creatures and plants they had never seen before, they focused on collecting and identifying, classifying, and organizing these organisms for museum collections.[7] Among all those discovered forms of life, some seemed to have special abilities that raised questions not just about what is normal but also about the organism's responses to the abnormal, or injury. Any organism can be injured or damaged, but a few eighteenth-century naturalists began to focus closely on those with the most obvious regenerative abilities. A crab or crayfish can regenerate a claw, for example, or as Aristotle had noted, a lizard can replace a missing tail. Other animals can repair injured skin, so that it looks perfectly normal again, or can recover from lost blood or fluids. These observations raised questions about whether it is only certain organisms or certain parts that can undergo such repair and, if so, which ones? And how and why does the repair occur?

In 1958, in an article "New (or Better?) Parts for Old," zoologist David Richmond Newth pointed to the enthusiastic interest in regeneration during the eighteenth century. He noted that "In 1768 the snails of France suffered an unprecedented assault. They were decapitated in their thousands by naturalists and others to find out whether it was true, as Italian Spallanzani had recently claimed, that they would then equip themselves with new heads." This popularization of empirical and experimental study of regenerating life attracted the attention of the French Enlightenment thinker François-Marie d'Arouet known as Voltaire. Newth obviously enjoys telling us that Voltaire "marveled briefly, saw at once that the loss and replacement of one's head presented serious problems for those who saw that structure as the seat of a unique 'spirit' or soul; and thought of the possible consequences of the experiment for man." On further reflection, Voltaire later noted that perhaps we might discover how to regenerate human heads and that for some people "the change could hardly be for the worse."[8] We probably each have our candidates for such replacements. All of this interest in the topic brought continued serious observing and imagining, and it also led to philosophizing about the causes of regeneration.

Though Aristotle did not discuss his causes in connection with regeneration, he likely would have invoked a final cause to lead the damaged individual organism to realize its potential and recover its appropriate form and function. By the eighteenth century, however, naturalists wanted something more in the way of explanation. Two sets of competing, and at times overlapping, ideas shaped thinking about normal generation and regeneration as a response to injury at least until the end of the nineteenth century with legacies today as well: vitalism versus materialism, and epigenesis versus preformation. Both took place in the context of a growing interest in natural philosophy, or what we today call science, as well as in philosophy more generally. Both drew on observations and also on a relatively new form of study through experience that is extended beyond what can be seen naturally without intervention called experimentation. Eighteenth-century natural philosophers realized that such experimentation could provide them with far more information than passive observation alone.

Issues of vitalism and materialism revolved around metaphysical questions about what exists in the world, as well as how best to understand it. For a materialist, living organisms as well as inanimate objects all consist of matter that is constantly in motion. There is no vital force or "stuff" that animates life and thereby separates living from nonliving nature. Vitalists, in contrast, did not see how to explain life without some special living "something." It might be a force or a substance or something even less concrete, but for vitalists there must be something unique to life to allow it to be living as a whole organized organism.

As historian Shirley Roe showed in 1981, debates about vitalism and materialism became entangled with ideas about epigenesis and preformation.[9] These are two competing interpretations of how individual organisms develop (though please note that epigenesis is not the same as what developmental biologists today call "epigenetics"). A traditional epigenesist, as Aristotle was, held that an individual living organism starts out in a disorganized, unformed way. Only gradually, over time, does the initially unformed substance develop and become organized through a process of generation to become formed as just the right kind of thing. A frog's egg becomes a frog, a chicken's egg a chick, for example. Anyone could see this organization taking place as they watched those chicken eggs or frog eggs develop, for example, so empirical observations reinforced their convictions that form and function emerge only gradually. Yet how does this occur, and how does the organism "know" how to develop in the right way to become the right kind of organism? Perhaps some vital factor drives the process, perhaps something like Aristotle's final cause leading toward the expected goal, or perhaps some other vital force or substance. They did not know what that vital entity might be, but many of these early epigenesists felt that it must be there.

In contrast, materialists, who rejected vitalism and insisted that all of life is made of matter in motion, disagreed. They understood that empirical observation suggested that form and function appear to emerge only gradually; they did not deny the evidence that they could also see as they watched the eggs develop before their eyes. Yet they insisted that this was just appearance, and if only we had better tools of observation like a better microscope, then we could see encapsulated within the sperm or ovum a miniature version of the organism's final form. After all, it must be there, because how else could each organism develop properly without some such guidance? These materialists held a preformationist view that living form must exist from the very beginning of any individual organism's life. They just did not know exactly how.

Thus, two sets of competing ideas driven by different underlying assumptions played out in complex ways. Each side held its views deeply, convinced that understanding the very nature of life was at stake.[10] It is worth trying to place ourselves at that time to appreciate the questions they asked and the kinds of answers they found acceptable. Today, if we see something we haven't seen before, for the vast majority of phenomena, we Google it and read all about it. Occasionally, we encounter something new like the "novel coronavirus" of 2020, known as COVID-19. Yet even here, despite the initial lack of much evidence about the particular novel virus, within a few months, scientists around the world pooled information and drew on knowledge of other related coronaviruses to inform understanding and suggest appropriate public health responses. What we do when we encounter something truly novel is to work to place it in the context of what we do already know about something similar, using methods we know have helped give answers for past questions. Imagine trying to do that in the context of relatively little past information and with available methods that are only being developed at the same time.

Intrepid naturalists of the eighteenth century nonetheless enthusiastically embraced the novelty and asked questions. In Europe, they reported their observations and suggested interpretations to their national scientific organizations, in most cases "royal" societies supported by their kings. These natural philosophers developed networks of colleagues who exchanged ideas, learned scientific practices, and sought to expand understanding by adding more explanations alongside descriptions. As they compared their ideas and debated possible interpretations, they could turn from description and classification of different kinds of life to explanation of living processes. One of the most fascinating of such processes was reproduction and development; and making progress in assessing competing ideas about epigenesis and preformation, or about vitalism and materialism, called for focusing on individual living organisms and the way they change over time. They needed expanded experimentation.

## CUTTING UP ANIMALS

In her *Catching Nature in the Act*, historian of science Mary Terrall gives us a fine study of what these curious folks did through her examination of the scientific network around the French naturalist René-Antoine Ferchault de Réaumur.[11] Her book explores the scientific working life of Réaumur and those with whom he interacted in person and by extensive correspondence by pointing out the people, practices, and publications through which his ideas were developed and conveyed. This generation of French naturalists were, Terrall shows, in fact working to catch nature in the act by observing and documenting the actions or processes of living organisms.

In Réaumur's case, she explains, he carried out the work with assistants who often lived at his estate to help with collecting as well as representing what they found. While admitting that it is better to be one's own artist, he apparently also realized that he was not particularly successful at drawing and therefore employed artists to capture images of what he observed. In fact, he was so grateful to his artists that he made sure to give them credit. In his will written in 1735, Réaumur gave the bulk of his estate (not including inherited property and a few smaller bequests) to his favorite artist "Mlle. du Moutier de Marsigli." He said, "I would like to be able to express to her all the gratitude that I owe her for the use she kindly made for me, with so much patience and steadfastness, of her talent for drawing. This is what made it possible for me to publish my Memoirs on the History of Insects and to continue this work [travail]. Whatever inclination I might have had for this work [ouvrage], I would have despaired of finishing it and I would have abandoned it on account of the time I would have lost had I been obliged to continue to have ordinary illustrators work under my eyes."[12] Apparently she had what Réaumur felt was a brilliant instinct for knowing just what to do and how to do it, so that she did not need constant instruction. It is unfortunate that her life story remains hidden in history.

Réaumur is not a well-known figure in the history of science, but Terrall certainly makes the case that he should be. A member of a wealthy family, he had the opportunity to follow his curiosity, which carried him to mathematics, physics, and also natural history. His accomplishments include developing a process for making steel based on his knowledge of metals, and he provided a simple and clear approach for measuring temperature with what became known as the Réaumur scale. Add his work in mathematics, and his wide-ranging curiosity reflects an immensely creative mind supported by the resources to follow his interests.

That lively curiosity extended to explorations of how living processes work, where he observed what he could see and devised experiments to extend those observations beyond what was immediately visible. Some of his elegant experiments explored whether simple organisms such as various types of insects can spontaneously generate, for example. He determined that they do not, and this realization reinforced his ideas about normal development and regeneration. He carried out extensive studies of insects, including detailed behavioral studies that have led to his being considered an early contributor to ethology. Réaumur was an active participant in the French Royal Academy of Sciences and also became a Fellow of the Royal Society of London, so he was well-connected in ways that enhanced his research, as Terrall demonstrates.

Among his many studies, Réaumur presented a paper in 1712 to the French Academy, suggesting that animals have a special ability to replace lost parts precisely where and when they need them. In particular, he was referring to the crayfish's ability to replace just those parts that they tended to lose through injury, and noted that "Nature gives back to the animal precisely and only that which it has lost" (fig. 2.1). In other places he established that various echinoderms (like starfish) can also replace their injured arms.[13]

Réaumur began by noting that many people lived near the shores of rivers or the sea and therefore had ample opportunity to observe all these various animals but had nonetheless missed something. These people rejected the idea that the animals in the waterways around them could regenerate missing parts as just "fables." Not true, Réaumur insisted, and we have only to look carefully to see the evidence. Some crabs or crustaceans have claws or limbs of different size, for example. Just

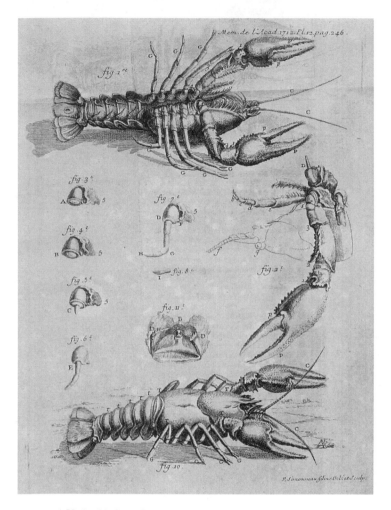

**FIGURE 2.1** | Plate 12 from René-Antoine Ferchault de Réaumur, "Sur les Diverses Reproductions qui se font dans les Ecrevisse, les Omars, les Crabes, etc. Et entr'autres sur celles de leurs Jambes et de leurs Écailles," *Memoires de l'Academie Royale des Sciences* 1712: 246.

look and you can see these parts getting injured or lost, and then you can see the organisms develop the missing parts. All it takes is careful observation, extending over time and across different individuals. Experimental intervention extends what is possible to observe.

Réaumur therefore asked what would happen if he intervened to cut off the legs of crabs and placed the crabs back into the sea. He wanted to determine whether they could regenerate new legs while living in the same environment in which they normally live, an experiment that did not succeed. Yet he did succeed in a similar experiment with crayfish. Cut off parts, and the crayfish can regenerate those parts and become whole again. Development biologists Dorothy M. Skinner and John S. Cook pointed to this phenomenon and to the timeless beauty of Réaumur's experiments for modern crustacean biologists in their 1991 essay praising the ingenious approaches of this "remarkable man."[14]

Réaumur's extended observations and experimental results raised questions about how and why regeneration occurs, as well as the extent to which regeneration follows the typical processes of generation. A materialist could explain the phenomenon by pointing to an intrinsic, and perhaps preformed or predetermined, capacity to regenerate parts. Notably, those parts that seemed to regenerate also seemed to be those most likely to undergo damage, so perhaps that was key. Or perhaps something about the whole organism invoked some vital principles to guide replacement in just the right way. The interpretations remained open, and debate continued. Increasingly, biologists thought of the organism as an organized whole system made up of parts with integrity and individuality that distinguished it from the surrounding environment, to which it nonetheless responded.

Réaumur reportedly served as something of a role model for the younger naturalist Abraham Trembley, who also shared his discoveries of regenerating polyps through a series of letters with Réaumur. Both carried out a wide range of naturalist studies, while Trembley focused particularly closely on phenomena and questions about regeneration. Réaumur came from relatively wealthy lower nobility in France, and Trembley's family was a long-established and well-respected family in Geneva, so they both had the privileges of education and leisure to carry out exploration. Trembley began his education with mathematics but then took up his very original studies of what he initially called polyps and later identified as hydra. These began while he was employed as a tutor from 1740 to 1744 and living with Count Bentick of The Hague. In fact, he may have been stimulated to pursue particular studies in part because he could share them with his young pupils. Biologists/historians Howard Lenhoff and Sylvia Lenhoff report that Trembley would visit the ditches on the Count's estate and collect glass jars full of things to study. The Lenhoffs point to Trembley's having recounted that he found that these jars "populated with little creatures" provided "good company with which to relax from more serious occupations."[15]

In describing Trembley's experiments, the Lenhoffs point to his precise description of his methods in the 1740s, inviting others to carry out similar experiments. He described using scissors to cut the polyp across its body.[16] His first discoveries established, perhaps by accident, that hydra are attracted to light. From there, he began close further observations that raised additional questions. Why do different individuals have different numbers of arm, or tentacles, for example? How might he add experimental ways to discover under what conditions those arms can arise?

Howard Lenhoff and Sylvia Lenhoff persuasively argued that Trembley should be thought of as an organismal biologist who saw the living organisms as systems. In studying living organisms, Trembley looked at the interacting whole, its structure, function, and behavior. The Lenhoffs also wanted to convey that in an era when many naturalists, including Réaumur, explored widely across diverse forms, Trembley specialized in studies of what soon came to be called hydra. Early in his studies, he established that hydra are animals rather than plants as many had previously thought. He then began extensive observation and experimentation of these fascinating creatures. Turning an individual inside out, cutting off parts, sticking different parts together: all these experimental manipulations produced enticing results. The organisms kept living, and they quite readily acclimated to their new conditions by restoring their normal structure and function. Trembley clearly found the hydra's abilities to regenerate important for understanding fundamental processes of life.

The hydra was a new and fascinating organism for Trembley, and his accounts made it fascinating to many others as well. What would happen if he cut one in half, he asked, and then proceeded to make cuts in various directions? Each part grew into a whole organism, replacing the missing parts and functioning in apparently normal ways as a result. He offered detailed descriptions of precisely what happens with each different kind of cut. Cutting just the head led to a two-headed hydra, with further splits producing four or eight heads. He then turned the hydra inside out to ask just how the organism responds. If left alone, it reverted back to its original shape. But if he prevented that from happening, then it adapted to the new circumstances to become a normal-looking animal. As he reported these observations, he enticed others to conduct similar experiments.

Trembley noted that while earlier speculations about regeneration in animals had imagined that an organism cut in half might be able to join the parts back together, or it might be possible to reattach a severed body part, here was a case where the organism had parts cut off and replaced them entirely. The animal seemed able to generate its parts out of nothing. "Here," he said, "is nature going further than our fancies."[17] In fact, he concluded that not only had the organism replaced its parts, but it was apparently just the same as an organism that had developed normally: "I found no difference between the polyp that developed from the second [cut] part and a polyp that had never been cut up."[18] Given later emphasis on studies of regeneration in the flatworms called planarians as well as hydra, it is amusing that Trembley included those flatworms among the many things that his hydra were apparently happy to eat (fig. 2.2).

This is quite remarkable research and reminiscent of Aristotle in demonstrating relentless curiosity and unwillingness to settle for accepted theory when Trembley's own empirical

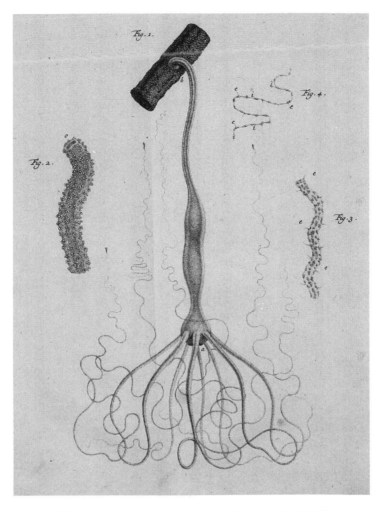

FIGURE 2.2 | Plate 5 from Abraham Trembley, *Mémoirs pour servir à l'histoire d'un genre de polypes d'eau douce, à bras en forme de cornes.* Leiden: Verbeck, 1744.

observations showed him otherwise. When his experience
showed him that one small piece of a hydra could regenerate
a whole organism that looked normal, he believed his results.
The fact that this observation did not fit with the preformationist idea that the organism must be present somehow from the
beginning just told him that we need a new or different theory,
and perhaps the rival epigenetic theory would do. The organism seemed able to develop gradually. Yet that did not mean
that preformationists could not offer alternative accounts consistent with materialism and somehow building in auxiliary
capacities for the organism. Trembley's observations stimulated lively discussions. As the Lenhoffs noted after their extensive review of all his writings, Trembley himself did not seem
interested in the philosophical interpretations and even "har-
bored a distaste for theories in general."[19]

As Trembley reported on his results, Réaumur continued
his own studies cutting hydra into a variety of different pieces
in different orientations, which led to further observations and
ideas to share with Trembley. While Trembley had become
a specialist with hydra, Réaumur also carried the technique
to other animals. He cut earthworms into pieces to see what
happened. And he urged others to join him in doing the same
with marine organisms, especially starfish. All these organisms
could regenerate so that they restored their original form and
function. Perhaps such capabilities held only for certain types
of organisms, Réaumur suggested. Charles Bonnet and Lazzaro Spallanzani joined the discussions with their own studies.

Like Trembley, Bonnet came from Geneva. He worked as a
lawyer but took pleasure in studying natural history as well and

was also inspired by Réaumur's work. In 1741, Bonnet began to study regeneration. Among other observations, Bonnet looked at worms, cutting them into parts to observe the resulting restoration. Even cutting a worm into fourteen pieces could produce fourteen new worms, he found, each apparently "correct" in its form and function. Similarly, Bonnet concluded that each hydra had an integrity and individuality that defined it. Chopping off a head (or a part) simply released what was, in effect, its suppressed self or organizing principles. For Bonnet, each organism contained the "germs" necessary to restore parts. His type of vitalism and epigenesis went along together.[20] In fact, historian Shirley Roe shows that Bonnet's regeneration experiments and interpretations convinced the Swiss naturalist Albrecht von Haller to change his mind about how development occurs. As a materialist, Haller had held a preformationist view. The ability of organisms to regenerate, and Bonnet's accounts in particular, caused Haller's conversion to the gradual development of epigenesis while he also sought to remain a materialist and to find the natural laws guiding the processes.[21]

Years later, in 1768, the Italian Lazzaro Spallanzani reported on similar experiments with earthworms, slugs, snails, tadpoles, salamanders, and frogs. Spallanzani also came from a wealthy family in circumstances that allowed him to explore natural history alongside other sciences. His studies included mathematics, physics, and law, and he earned a doctorate in philosophy. He became ordained as a Roman Catholic priest, which afforded him some financial support for his studies, though he clearly invested his efforts in exploring nature.

One set of studies involved cutting an individual into parts to determine whether the dissociated parts produced more individuals. Another study involved cutting off only smaller parts to determine whether the individual would regenerate the missing part. Spallanzani summarized his results of numerous observations in a 102-page "letter" transmitted to the Royal Society of London and apparently intended it as a prelude to a larger publication considering reproduction generally. In fact, it reports primarily on regeneration as a response to damage in organisms. Fortunately, the Society's secretary, Matthew Maty, quickly translated it into English and thereby made it available in 1769 to a much wider audience.[22]

Describing what experimental manipulations he had made, including how and where he cut the animals, Spallanzani reported also on how they responded. How did the structure develop, how did the blood flow, in what direction did the changes occur, and many other details. He noted, for example, that the blood in the regrowing tail of a tadpole circulates differently from that in the original tail. This led him to discuss the "new organization" of the regenerating organism. Yet it is clear that at least in some important respects, Spallanzani held that the organism (or at least some "germ" providing the basis for forming a new organism) is somehow present already in the egg. The egg then follows a process of developing from that initial form. This sounds like preformationist thinking. At times, Spallanzani held such a position; at others he sounded more sympathetic to an epigenetic position. In a series of letters with Bonnet, for example, he wondered whether the frog with the severed tail might actually be reshaping the material it already had rather than generating a new tail.[23] He felt it important to follow the observed evidence and assess theoretical interpretations based on what he saw.

While each of these men carried out his work in the context of his commitments to materialism and preformationism, or epigenesis and vitalism, they did not resolve the fundamental debates about the nature of life. Does regeneration occur because of some kind of preformed "auxiliary" parts in organisms likely to be injured, perhaps? Or is there more of an gradual epigenetic adaptive response, perhaps guided by some internal vital principle?

## EXPERIMENTAL EMBRYOLOGY

In fact, the questions about regeneration remained much the same until the late nineteenth century, which brought a number of changes for the life sciences. The term "biology" caught on only around 1900, despite having been introduced a century earlier.[24] The emphasis on observation and empirical evidence continued but was complemented by increasingly interventionist experimentation that went beyond the kinds of relatively simple manipulations we have been discussing. While the eighteenth century called for extended experience through manipulation of observations, which they called experiments, the end of the nineteenth century version required additional assumptions about whether the observer was actually observing something normal, as well as an emphasis on developing theories and the role of hypothesis testing.

German embryologist Wilhelm Roux promoted this bolder experimental turn, as it has been called, most vociferously. He presented his ideas especially through the journal he edited on the developmental mechanisms of organisms, called the *Archiv für Entwickelungsmechanik der Organismen*. His introduction to the first volume in 1894 included a sort of manifesto. Roux edited the journal until 1923.[25] The journal has continued through a series of title revisions and is now the journal for *Development Genes and Evolution*. Roux's opening declaration of the need for what he argued was a new experimental biology attracted considerable attention. Across the Atlantic, biologists at the Marine Biological Laboratory in Woods Hole, Massachusetts, heard about the ideas when William Morton Wheeler offered a translation of Roux's essay and in 1894 led a lively discussion in what later came to be known as the public and popular Friday Evening Lectures.[26]

Roux had attracted wide attention, in part because of the rather energetic way he expressed his ideas and dismissed alternatives, but also for offering a definite theoretical interpretation of how development occurs. With an interest in embryological development, he saw experimental manipulation and the production of abnormal conditions as a way to investigate a process that was otherwise invisible. He poked and prodded various frog embryos, and then in 1888 he published the results of an experiment for which he declared dramatic interpretations.

Frog eggs provided a very useful subject for studying development. They are relatively large, visible even without a microscope, outside the body, readily available in the right seasons and in the right places, and with different species to facilitate making comparisons. Historians have provided considerable detail and perspective on the range of Roux's frog studies, so here we focus on Roux himself and his ideas about development.[27]

Roux started with the idea that development of an organism occurs through a process something like that involved in producing a mosaic. Cells serve as individual tiles for the mosaic, or the fundamental building units, and their arrangement turns the parts into a whole. Somehow, Roux imagined, the fertilized egg cell contains what it needs to progress through cell divisions and specialization of cells over time. The process itself appears epigenetic, in that traditional sense that the form of the individual organism emerges only gradually. Yet the embryo and its cells are clearly material and are the products of mechanistic development, not guided by vital forces or entities. Perhaps the fertilized egg might be predetermined in part, but Roux did not put his theories in the traditional preformationist terms in which the form itself was encapsulated from the beginning. Instead, he sought to set up a new paradigm, with mechanistic studies of materialist life processes driven through specialization of cells. Somehow.

Roux, like his fellow German naturalist August Weismann, focused on the nucleus as the source of determination for each of the cellular parts. A number of researchers were discovering that the nucleus had special structures, in the form of chromosomes, which raised new questions about its role in the cell and in development. Roux became convinced that the chromosomes must contain inherited material that is divided during the cell divisions so that the original molecular material becomes distributed in different cells. This might explain how cells specialize during development, he argued. Surely the nuclear structure must provide the key for guiding development and organization of the complex organism. Weismann developed similar ideas in much more detail, and with some differences from Roux's ideas, but the two are often lumped together with the "Roux-Weismann" or "Weismann-Roux" theory, because they both held the chromosomes as central for producing life through new developing individual forms.[28]

After pursuing various other experiments, in 1888 Roux published the results of the one that gained him the most attention and generated the most controversy about the proper way to do science. His methods started with fertilized frog eggs. Normally, the initial cell divides into two cells, then into four, into eight, and so on, though not quite as evenly after the first divisions. Roux hypothesized that at each division, the cells that resulted had become different. They had become separate pieces of a mosaic. Thus, each of the two blastomeres, as they were called, had its own individuality. In particular, the division of different nuclear materials into each of the different cells might be what determined just how each cell would differentiate. This division followed Roux's complicated theory of nuclear materials acting through a "Kampf der Theile" or struggle of parts. To test this hypothesis, he killed one of the two cells after the first division. He did this by poking it with a hot needle until it no longer continued to divide or to act like a cell in any way. He predicted that the result would be a half embryo, where the missing cell would not develop and the embryo would not be able to replace it (fig. 2.3).[29]

Sure enough, one cell developed and the other did not. Sure

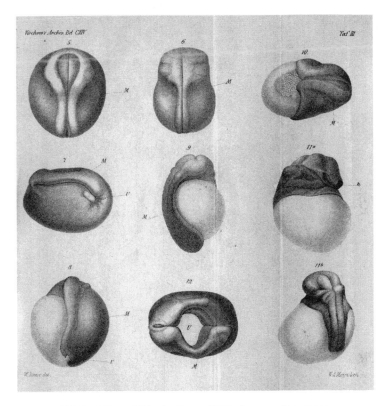

FIGURE 2.3 | Plate 3 from Wilhelm Roux, "Beiträge zur Entwickelungs-mechanik des Embryo. Über die künstliche Hervorbringung halber Embryonen durch Zerstörung einer der beiden ersten Furchungskugeln, sowie über die Nachentwickelung der fehlenden Körperhälfte," *Vir-chows Archiv für Pathologische Anatomie und Physiologie und für Klinische Medizin* 114 (1888): 113-53.

enough, the result looked to Roux like a mosaic. He declared his hypothesis as confirmed. Each cell contains an ability to self-organize in the appropriate way, and the resulting whole organism is built up out of the parts and not guided by some kind of mysterious, vitalistic wisdom of the whole, according to Roux. In fact, Roux later admitted that the story was a bit more complicated, and in some cases the "killed" cell actually started to develop, but he did not regard this as disconfirming his hypothesis.

Hans Driesch disagreed. And he disagreed in ways that created a great deal of debate and great annoyance for Roux. These two German embryologists had very different ideas about how to do science as well as about the structure and nature of the living world. Driesch started out along the path that Roux had taken. He was fascinated by Roux's results but also felt that he could do a better and cleaner experiment. Roux had poked one of the two cells to kill it, but that left the inert mass of material sitting there and perhaps exerting some influence. Driesch asked whether he might use sea urchins instead of frogs, since he knew that it was possible to shake its embryo at the two-cell stage and to separate the cells completely. And so he did. He shook, the cells separated, and he explained that when he left for the night he expected to return the next morning and find the same result that Roux had found with frogs. In that case, he should have found one live and one dead cell. But no, he found that the two separated cells had each kept developing and they then went on to become two small pluteus larvae (fig. 2.4).[30]

Driesch had used an experimental approach, just as Roux required. He got different results, and that happens in science. But to Roux's great annoyance, Driesch also developed a very different interpretation of the results and what they meant for embryonic development. For Driesch, the result showed that development does not occur in any mosaic-like way. Nor was there any evidence that nuclear material divided into the different cells. Driesch did not see cell division as separating the

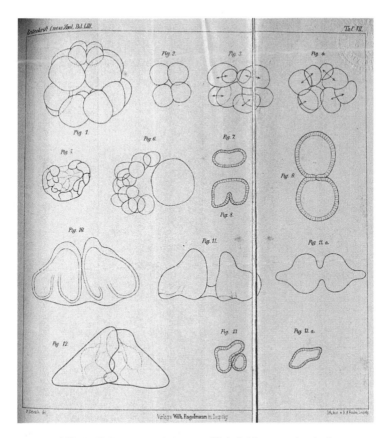

FIGURE 2.4 | Plate 7 from Hans Driesch, "Entwicklungsmechanische Studien: I. Der Werthe der beiden ersten Furchungszellen in der Echinogdermenentwicklung. Experimentelle Erzeugung von Theil- und Doppelbildungen. II. Über die Beziehungen des Lichtez zur ersten Etappe der thierischen Form-bildung," *Zeitschrift für wissenschaftliche Zoologie* 53 (1891): 160–84.

cells into different parts that then develop according to their own self-organization. Instead, Driesch argued that there must be some process he called "organismal regulation." By that he meant that something about the whole organism guides the direction of the parts. Eventually, Driesch adopted an avowedly vitalistic approach, then gave up scientific research and took up the philosophical study of metaphysics, or an account of what exists in the world. He called the early blastomeres "totipotent," because they retain the capacity to develop into the whole organism.[31]

Arguments erupted about who was right. To what extent are cells self-determined and self-organizing? To what extent and in what ways are cells and the developing organism able to respond to damage? Is the developing organism an integrated whole that is organized as a whole and follows some process to regulate it? Or perhaps these views, which have been presented as the neat dichotomies of materialism versus vitalism and preformation versus epigenesis, are not quite so distinct as they had been presented. Surely it was necessary to gather further evidence to make progress in explaining development. In particular, it was definitely necessary to carry out a great deal more experimentation to understand the role of chromosomal division and assess Roux's claim that cell division divides nuclear material into different parts that distribute into different cells.

In addition, Roux needed to explain how Driesch got his results since each of the separated cells should not have been able to develop into a whole organism in Roux's interpretation. Fortunately for Roux, his methodological approach to science allowed him to add "auxiliary hypotheses" if things did not turn out quite as expected. He therefore invoked the concept of "postgeneration," arguing that development is still mosaic. Somehow, the nucleus contains a set (or more than one set) of reserve determinants that can be brought into play in the case of injury. The result is still a mosaic, just a secondary one through regeneration. A set of reserve materials kick in. Thus, central to these debates lay questions about regeneration.

Roux did not start studies of regeneration, obviously, but his experiment and Driesch's different results reframed questions about what regeneration really does. How does an organism regenerate a part or a whole? Roux and Driesch joined ongoing discussions concerning interpretations of regeneration and began to point to regeneration as a way to use the organism's response to abnormal conditions to illuminate understanding of normal developmental processes.

Historian of science Frederick B. Churchill has noted what he termed "this apparent chaos of activity" at the end of the nineteenth century in regeneration studies. Building on the rich history of observations from the previous century, and in the context of Darwin's ideas of evolution and advancing techniques for experimentation, "droves of experimentalists and anatomists turned to their scalpels, scissors, and heated needles; they shook, ligated, and compressed; they altered the orientation of the egg to gravity and changed the chemical composition of the seawater," in order to understand the fascinating phenomenon of regeneration.[32]

Within the busyness that Churchill identifies, the German anatomist from Rostock, Dietrich Barfurth, stands out. Between 1891 and 1916, Barfurth produced twenty-three reviews of studies of regeneration. His kind of review article served to summarize in published form the work by a number of researchers who might not be publishing results formally, as well as to document changing questions and methods. Barfurth's reviews appeared in the journal *Ergebnisse der Anatomie und Entwickelungsgeschichte* and, as Churchill notes, reflect the author's particular interpretive preferences. The reviews also show that by the second half of the nineteenth century, it was frogs and their eggs that gave their lives for science. The period also saw publication of several volumes devoted to study of regeneration, including *Regeneration* by German physician Paul Fraisse in 1885. Though interacting with ideas of the day based on microscopic studies, Fraisse's contribution was more summary of select observations than interpretation. It received relatively little attention, as the flow of regeneration studies continued.

In contrast, Barfurth's reviews offer valuable insights and reflect an excitement about discovery such that it is worth looking more closely at Churchill's account of those reviews and their place within the move to increasingly experimental approaches to biology. These reviews show the role of observations in stimulating interpretation, which then guides further experimental exploration to examine the theories. They show the development of a modern experimental approach to biology while also showing that regeneration had become increasingly regarded as part of the ongoing process of generation. Development of individual organisms occurs, starting with fertilized eggs and proceeding through a sequence of cell divisions, cell specializations, and organization of the parts into the whole functioning organism. Interrupting that process, or damaging some part of the resulting individual evoked a regenerative response in some cases, but not all. When, how, and why became questions about embryology and development as well as about the special phenomenon of regeneration.

By looking closely at the reports of studies and their interpretations over the first decade of the twenty-five years of reviews by one author, Churchill shows just how rich exploration of regeneration was around the turn of the twentieth century. It becomes clear why Thomas Hunt Morgan—known for his 1933 Nobel Prize–winning work on chromosomal heredity in *Drosophila* genetics much more than for his studies of regeneration—took up the topic as a particularly useful way to explore experimental embryology.

## THOMAS HUNT MORGAN

We focus on Thomas Hunt Morgan's observations and his reasoning in much more detail in chapter 3 in connection with his important 1901 book *Regeneration*. But it is useful here to see him as providing an excellent summary of the diversity of studies that preceded 1901. Throughout his life, Morgan made clear that he was intrigued most by the phenomenon of development of individual organisms and fascinated by their ability to regenerate parts in some conditions but not others. Historian of science Mary Sunderland has argued that Morgan saw the study of regeneration as an excellent window into development, while historian and philosopher of science Jane Maienschein suggests that Morgan's study of regeneration was a way for him to explore what allows complex organisms to work as whole apparently organized units or systems.[33] Without giving in to vitalistic accounts that invoked a special vital something, biologists could still acknowledge something more than just Roux's mosaics working independently. How does development work? Morgan asked.[34]

As Morgan wrote in his 1901 summary volume: "It can be shown, I think, with some probability that the forming organism is of such a kind that we can better understand its action when we consider it as a whole and not simply as the sum of a vast number of smaller elements." That is, "the properties of the organism are connected with its whole organization and are not simply those of its individual cells, or lower units."[35] We return to this idea in the next chapter with the move to development and regeneration in terms of mechanisms.

# 3    Mechanisms of Regeneration

The story so far has shown the ways researchers understood and wove regeneration into the history of biology up to the twentieth century. Throughout the millennia spanning between Aristotle and Thomas Hunt Morgan, observers of nature recorded that organisms when injured possess the capacity, to varying degrees, to recreate what they had lost. The major concepts tied to generation in the Enlightenment, like epigenesis and preformation, as well as the major worldviews through which scientists perceived the world around them, like vitalism and materialism, were part and parcel of making sense of regeneration. As experimental scientists in the late nineteenth century such as Wilhelm Roux turned toward explaining natural phenomena in terms of internal causes and mechanisms, they began to reenvision regeneration as a fundamental process of living systems. Furthermore, they felt that understanding regeneration required the latest experimental techniques and new ways of understanding organisms' development, including taking control and manipulating the organisms as well as drawing on more systems-oriented approaches. This new perspective came to fruition in the early twentieth century.

We could follow a number of different scientists through their contributions to understanding regeneration, but our aim is to provide a foundation for understanding modern thinking. Therefore, we have chosen to look closely at three scientists whose work overlapped and intersected, with instructive results. These scientists, Thomas Hunt Morgan, Jacques Loeb, and Charles Manning Child, asked many of the same questions about regeneration but took different approaches toward investigation. They all embraced a systems-based approach to understanding their research organisms and regeneration. Two of them wrote books entitled simply *Regeneration*, while the other wrote volumes with regeneration as a central theme. They also all worked in the United States at a lively time when American biology was emerging as a professional study of living systems.

About the same time in 1901 that Morgan was laying out his summary of work to date on regeneration, Loeb also saw regeneration as a central phenomenon that would help illuminate understanding about how life works. Loeb sought to demonstrate that life can be understood in purely physico-mechanical terms, in particular as the actions of fluids responding to changing conditions within an organism and as a quantitative function of the mass of the organism's parts. Changing external conditions can injure or damage the organism and thereby cause internal changes, and Loeb wanted to understand how and why that occurs and how the organism responds. Furthermore, for Loeb this was not just a matter of intellectual curiosity or simply wanting to know how life works. He really wanted to go farther and control life. If we can understand how regeneration works to restore function, he urged, we can use that knowledge to improve and extend life in ways that over a century later led to the field called synthetic biology, in which researchers seek to synthetize life in their laboratories.

At the same time that Morgan and Loeb were busily examining regeneration, Child was exploring patterns of organization in biological systems and asking what mechanisms shape organization in development and in regeneration after injury. His idea of internal gradients of influence evoked considerable discussion and disagreements concerning how developmental processes work.

These three individuals interacted intellectually and sometimes personally, addressed overlapping questions, and yet developed different methods and approaches for studying and explaining their results. Looking at all three closely shows a great deal about how biologists approached organisms in that time of newfound experimentation and interpretation based on materialism and rejecting all vitalism. It also shows how those biologists came to see regeneration as a central part of living systems rather than as a specialized exotic response that only occurs occasionally when systems respond to damage. Throughout this chapter we follow the efforts of these biologists in some detail during the period of 1900 into the 1920s in order to tease out their conceptual and methodological contributions to regeneration thinking that is foundational to our modern perspectives.

## CONTEXT OF UNITED STATES BIOLOGY AROUND 1900

Historians of biology have written a great deal that very instructively shows the self-conscious efforts by Americans first to join international discussions about biology individually, then to establish an American tradition in biology with solid institutional grounding. This effort required creating an infrastructure to support research and education while offering paid professional professorships. Since the generation who became biologists around 1900 and after came largely from middle-class backgrounds, they needed employment and depended on teaching positions at the growing number of American universities. The Johns Hopkins University, modeled on British education systems, provided a PhD program with graduate fellowships to train biologists. Then the fortunate few found positions at what was becoming a set of research universities that supported laboratory work, such as Columbia, Yale, Harvard, Stanford, and the University of Chicago. The Marine Biological Laboratory and, later, other research stations provided a place for researchers to gather and to learn from each other, typically in the summers, since each might be one of just a few biologists in their home institutions but they could all find a community of like-minded others at these summer stations. The American biologists were typically pragmatists, grounded in empirical observations while also well aware of the larger international context of theory and expanding technologies.[1]

## THOMAS HUNT MORGAN

Morgan gives us a starting point for understanding this American tradition, and in particular the place of regeneration studies within it, with his 1901 volume *Regeneration*. By that point, he had already published twenty-one articles with the term regeneration in the title. Most of those articles offered results of empirical studies, looking at injuries and responses in such diverse organisms as several kinds of worms, planarians, hydra, hermit crabs, fish, sea urchins, and the trumpet-shaped ciliated small animals called stentors. Other lectures and publications examined his assessment of different explanations that had been offered by others, especially in Germany.

By 1901, Morgan had already developed a working strategy with his first book on *The Development of the Frog's Egg: An Introduction to Experimental Embryology*.[2] He carried out experiments, wrote scholarly articles and published his descriptions, offered lectures that laid out his thinking about the descriptions in the context of other work, got feedback from others, then summarized all of it in a book. He took precisely this same approach with regeneration. Through the 1890s, he had also developed a pattern of carrying out research on whatever marine organisms were available and seemed useful to address his questions. He did this during summers in the small town of Woods Hole, Massachusetts, at the southern tip of Cape Cod. There he was part of a lively community who gathered regularly to discuss ideas and discoveries both at the Marine Biological Laboratory (MBL) and at Morgan's house and small farm on Buzzards Bay Avenue.

While historians of science have written extensively about Morgan's research program studying the roles of chromosomes in fruit fly heredity that led to his 1933 Nobel Prize, relatively few have paid attention to his work on regeneration and development. To Morgan's main biographer Garland E. Allen as well as to other historians of science writing when genetics had become dominant for biology, Morgan's discussions of regeneration and of embryology more generally simply did not seem connected with what they saw as his more important study of genetics. In fact, many of Morgan's contemporaries and historians interpreted that even Morgan's book in 1934 entitled *Embryology and Genetics* did not really connect those two fields directly. Yet Morgan felt that he had done so, noting that he had included discussions of both embryology and of genetics, and had in fact raised substantial questions about the efficacy of genes and the ways that genes can interact with the rest of the organism.[3] Given the lack of attention to Morgan's regeneration work and given how much effort he spent on this work and how important he saw it, we find it valuable to look closely at Morgan's ideas about regeneration as a way of understanding his broader thinking about living organisms.

Study of regeneration by 1901 and well after remained largely focused on morphological studies of what happens to an organism's structures after an individual receives an injury rather than physiological discussions of the impact on function. Though Weismann and Roux had suggested connections with heredity and chromosomes, the leading causal mechanisms put forth as explanations of regeneration did not include the new post-1900 science of genetics until the second half of the twentieth century. Morgan himself did see the importance of heredity for biology, but he did not see ways to connect the genes themselves with developmental phenomena such as regeneration. Instead of his better-known studies of the fruit fly *Drosophila*, Morgan insisted throughout his life that his favorite organism for study was the sea squirt *Ciona*, and he also clearly enjoyed studying planarians and their regeneration. He retained an interest in regeneration as a process of development rather than a process connected with inheritance.

Morgan was born in Kentucky in 1866. His path to becoming a professional biologist began with a PhD from the Johns Hopkins University in 1890, directed by zoologist William Keith Brooks. The job market for biology professors was minimal in the late nineteenth-century United States, so Morgan was fortunate to receive a position at Bryn Mawr starting the next year, in 1891. There, he replaced cell biologist Edmund Beecher Wilson, who had also graduated from Johns Hopkins and was moving on to a professorship at Columbia University. Founded in 1885, Bryn Mawr had quickly become one of the leading women's colleges in the United States, with the progressive mission to educate women across all fields, including the sciences. Morgan stayed there until 1904, when he again followed Wilson and joined him at Columbia University. He remained there until 1928 when he moved to California Institute of Technology. In 1904 he married former Bryn Mawr student Lilian Vaughan Sampson, who continued to carry out embryological research at Columbia and at the MBL.[4]

At the MBL in the summers, Lilian wanted to work in the lab with Morgan, including on studies of regeneration, but

FIGURE 3.1 | Thomas Hunt Morgan and others having a picnic around Woods Hole, MA. Image courtesy of the Marine Biological Laboratory Archives. https://hdl.handle.net/1912/21003

the couple had four reportedly very lively and energetic children. Fortunately, Morgan's mother joined them for summers in Woods Hole to help with the children. One picture of a picnic at the beach (fig. 3.1) shows Morgan's mother seated on a rock in her full black dress. Morgan dedicated *Regeneration* "To My Mother."

Presumably because of their connections at the MBL every summer, Morgan received an invitation from Wilson and his Columbia colleague and department head Henry Fairfield Osborn to present a series of lectures at Columbia. In January 1900, Morgan reports that he went to New York and gave five lectures on "Regeneration and Experimental Embryology." His introductory comments show that he had come to hold an idea of regeneration as an important normal part of development. His research started with observations and experimental studies of eggs and embryos as Roux and Driesch had done. He had also already become quite skeptical of explanations based on Roux's hypothetical actions of chromosomes and the nucleus as driving heredity and development. He also rejected Weismann's hypothesis that natural selection prepares some organisms to regenerate more easily, while others cannot regenerate at all. Neither of these hypotheses fit the phenomena as Morgan saw them. He then urged the importance of studying questions about life in a way that he considered scientific, which necessarily meant rejecting what he saw as such "unverifiable speculation."[5]

Morgan's commitment to careful, scientific study based on observations and hypothesis-driven experiments is abundantly clear in his 1901 book *Regeneration*. There Morgan conceived of organisms as dynamic, self-regulating systems. Within these systems exist tensions between different parts and processes, like growth and development, that the organism must constantly modulate throughout life. Injury "pulls" these tensions out of order, and this disruption evokes a systemic response to restore order through regeneration. Regeneration, to Morgan, is an extension of normal growth and development. We follow Morgan's structure and reasoning in order to examine both what he thought about the logic and rules guiding regeneration as well as what he regarded as the proper scientific approach to understanding life.

In his historical background, Morgan showed that he was very familiar with the contributions of Trembley, Bonnet, and

Spallanzani and had followed up by reproducing many of the observations himself. His overview of the breadth of knowledge about regeneration highlights how diverse the phenomenon actually is and what regeneration means. Salamanders can regenerate a number of different parts, while other animals, such as lizards, can only regenerate one or another part. Worms can replace a missing head but not the tail, depending on the circumstances. The "fresh-water planarians show remarkable powers of regeneration" (p. 9) with the ability to recover from a wide variety of cuts by producing new material of the right sort. In his opening section, Morgan drew on this broad range of studies on diverse organisms, showing the different ways in which different types of organisms respond to injury. In addition to regeneration in whole organisms, he pointed to the ability of eggs and embryos like Roux's and Driesch's to recover from loss of one cell as a type of regeneration.

Morgan pointed to, but rejected, accounts of what regeneration is given by Roux, Barfurth, and Weismann. Nor was he persuaded by Driesch's idea of regulation as guiding replacement. Instead, Morgan offered his own terminology and distinctions. *Restorative regeneration* concerns all those instances where regeneration occurs in response to injury. *Physiological regeneration* happens where replacement occurs—as when a bird molts its feathers, an elk sheds its antlers, or a crab sheds its shell and a new one grows as part of the natural cycles of life. Morgan proposed the term *epimorphosis* to refer to cases where the organism generates new material, which then undergoes formation into the new part to replace the old. A classic example of this is regeneration of a newt's tail following amputation, where the newt undergoes extensive cell proliferation in order to regrow its tail. Morgan proposed the term *morphalla-xis* to refer to cases in which the organism reshapes the cut or damaged surface into a new part without adding new material. For example, when a hydra is severed into pieces, each piece can reorganize into a new hydra without going through normal processes of cell division. This framework provided Morgan with a way to organize and discuss the empirical evidence and his interpretations in a rigorous and programmatic way.

Within this framework, Morgan sought answers to what causes regeneration and what conditions make it possible. He considered everything from external factors, including temperature, food, light, gravity, contact, and chemical changes in the environment, to internal factors, including polarity of head and tail, lateral organization, response of oblique surfaces, influence of internal organs, amount of new material, influence of old parts on the new, influence of the nucleus, and closing of cut-edges. These lists show what factors Morgan saw as influencing both whether regeneration happens and the way in which it occurs. He included his own observations while also reporting on the results of others, laid out so that anyone could follow. Though he referred to regeneration in plants, his focus remained on his various animals (figs. 3.2, 3.3, and 3.4). In keeping with his abhorrence of "unverifiable speculation," Morgan did not offer theories about how the factors worked in the regenerative process. Instead he pointed to weaknesses in his predecessors' speculations. He wanted to establish the boundaries of the phenomenon in as much detail and with as much clarity as possible.

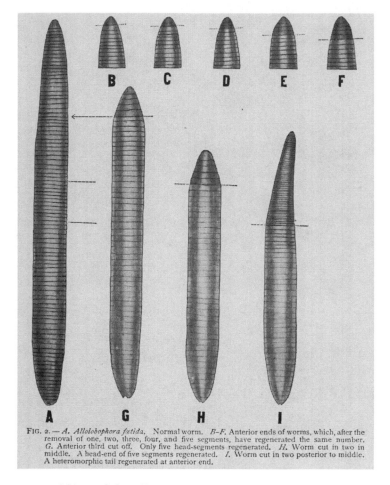

FIG. 2.— *A. Allolobophora fœtida.* Normal worm. *B–F.* Anterior ends of worms, which, after the removal of one, two, three, four, and five segments, have regenerated the same number. *G.* Anterior third cut off. Only five head-segments regenerated. *H.* Worm cut in two in middle. A head-end of five segments regenerated. *I.* Worm cut in two posterior to middle. A heteromorphic tail regenerated at anterior end.

FIGURE 3.2 | Figure 2 from Thomas Hunt Morgan, *Regeneration.* New York: Macmillan, 1901.

While recognizing the key questions of why regeneration happens in some organisms and some parts but not others, he rejected ideas such as Weismann's that evolution has adapted some organisms and parts because of their increased "liability to injury." It made sense, Weismann suggested, that evolution through natural selection would have led organisms to adapt to

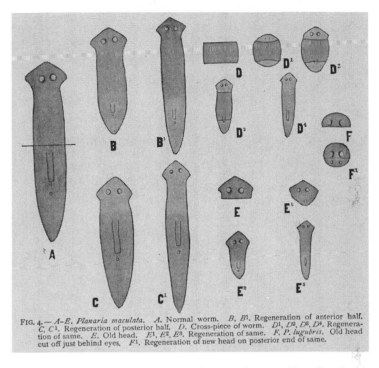

FIG. 4.— *A–E. Planaria maculata. A.* Normal worm. *B, B¹.* Regeneration of anterior half. *C, C¹.* Regeneration of posterior half. *D.* Cross-piece of worm. *D¹, D², D³, D⁴.* Regeneration of same. *E.* Old head. *E¹, E², E³.* Regeneration of same. *F. P. lugubris.* Old head cut off just behind eyes. *F¹.* Regeneration of new head on posterior end of same.

FIGURE 3.3 | Figure 4 from Thomas Hunt Morgan, *Regeneration.* New York: Macmillan, 1901.

the tendency to be injured. A lizard is likely to suffer an injured tail, crabs are likely to injure or lose their claws or legs, spiders lose legs, starfish lose arms, and so on. Weismann saw it as quite obvious that the ability to adapt to the losses would give the organisms an advantage, and therefore regeneration was an adaptation preserved through the Darwinian process of natural selection.

Morgan walked through Weismann's reasoning, as well as the arguments and results from other investigators, in detail. Yes, he concluded, the ability to regenerate is an advantage. Undeniably so. Yet sometimes these regenerating parts in the

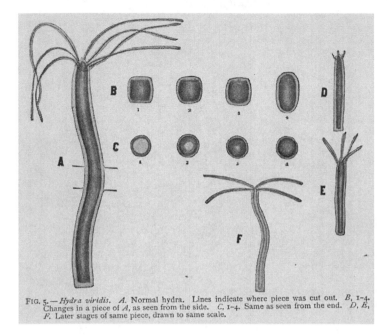

FIG. 5.—*Hydra viridis.* *A.* Normal hydra. Lines indicate where piece was cut out. *B,* 1-4. Changes in a piece of *A,* as seen from the side. *C,* 1-4. Same as seen from the end. *D, E, F.* Later stages of same piece, drawn to same scale.

**FIGURE 3.4** | Figure 5 from Thomas Hunt Morgan, *Regeneration.* New York: Macmillan, 1901.

experimental specimens take place where no injury occurs under normal conditions. Why does that count as evidence for adaptation rather than against this evolutionary interpretation? There are also clearly cases where regeneration is not adaptive, as when a head appears on the tail end of a planarian.

Morgan expanded his anti-adaptation perspective with further examples suggesting that regeneration can occur in a number of ways that respond to specific accidental conditions and are not "normal" or adaptive in any sense. Internal organs can respond to damage, or they can carry out a normal type of physiological regeneration, sometimes with abnormal results. The process is not perfect, which suggests a complex, interactive set of physiological responses by the organism and internal to the organism. Morgan considered the process in which individual cells separate from each other completely, as with twinning, and each develops into a whole organism through a type of regeneration. This phenomenon calls for some kind of internal coordination and guidance of the process, he concluded. Another example came with grafting parts from one organism onto another organism and watching the responses that suggested a kind of regeneration of a functioning whole organism. Morgan brought together all the different kinds of phenomena that could be considered regeneration and sought to lay out what was at issue in order to assess what it would take to provide explanations.

One question for all the cases concerned the source of new material: was it formed completely new in response to the injury (epimorphosis) or was existing material reformed into new parts (morphallaxis)? One approach involved looking at eggs and embryos as an example of regeneration, since so few cells were involved and he could easily observe the resulting changes. Roux's and Driesch's experiments on half embryos in frogs and sea urchins, respectively, played prominent roles in Morgan's thinking about regeneration in the embryo, and he carefully duplicated and responded to each of their works. While at first he focused on laying out details and observations, looking at the theories of others but rejecting them, eventually he turned to his own interpretive contributions that show his particular scientific approach and that were typical of his American colleagues. Morgan remained committed to an empirical approach, in the traditional sense, in that he urged beginning with observations, lots of observations, then a summary of observations.[6] Only after amassing a great number of observations should scientists attempt to develop interpretations.

Discussion of theories required Morgan to review contemporary German literature. Eduard Friedrich Wilhelm Pflüger had experimented with the effects of environmental conditions, such as gravity, on early cell division. So had Roux, Gustav Born, Driesch, and Morgan himself. The reports led to discussion of causes and effects in cell divisions and the mechanisms by which divisions both respond to and influence the rest of the embryo. Morgan suggested that these debates raised basic questions about how parts relate to the whole, with implications at the core of understanding regeneration and development. While Morgan rejected Driesch's vitalism, he did recognize the value of at least thinking about what might be meant by the "formative forces" Driesch and others invoked. Morgan simply concluded, "There must be assumed to exist in the egg an organization of such a kind that it can be divided and subdivided during the cleavage without thereby losing its primary character."[7]

Morgan turned from his point that there must be some organization in the egg to looking at what "method of action" that organization involves, particularly for regeneration. Again, he reviewed hypotheses offered by others, showing what is lacking in each and what questions each raises that would need to be answered before adopting that view. For instance, while Roux had suggested that some special "postgeneration" factor explains how the undamaged cell in his half embryo experiment recovered and developed, Morgan believed that regeneration involved the same factors as normal growth and development.

What, Morgan wanted to know, governs and regulates the process of regeneration, especially in light of the fact that it was an extension of normal development? One possibility was that the organism experiences a kind of tension throughout its body that exists throughout its lifetime. This tension was greater or lesser throughout the body, and the degree of tension present would correlate to the regenerative ability of that part of the organism. Disruptions of the tension would then evoke responses to restore order, so to speak. Planarians show this effect very well, Morgan noted, because splitting them and watching them regrow under different conditions shows limitations on when and where planarian can produce a head, for example, or not. The results can, "I think, be better appreciated if we suppose some sort of tension to be the influence at work."[8] Internal factors shaped by the internal organization therefore regulate growth such that the damaged organism becomes a new organized whole. Perhaps there is even a system of multiple tensions at work, or perhaps that idea is too vague, Morgan pondered. He suggested that an internal system of tensions might serve as a working hypothesis to be tested and refined.

Morgan reiterated that regeneration is not a special response to changing environmental conditions but, rather, an internal normal process of growth and development. Nor is regeneration an evolutionary adaptation to external conditions, even though the process may be useful. Instead, we must look within the individual organism and its parts, processes, and developments over time. Morgan sought to leave the subject not with ultimate conclusions or a big theory but with suggestions for ways to move forward with interpretations not driven by assumptions that could not or had not been tested. He opened the door for others to join the discussion. And they did, starting with Jacques Loeb. Loeb also looked to internal factors but was much more persuaded of the influence of external environmental conditions interacting with internal organization. And Loeb demanded, above all, a mechanistic view of life.

## JACQUES LOEB

Jacques Loeb was born in Germany in 1859. Originally named Isaak, he changed his name at age twenty after both parents died and he moved away from his Jewish family and on to an education in the humanities and then medicine. His wide reading served him well in his later theoretical work, in which he often called on philosophical perspectives. His father's sympathies for the French took him to French as well as German literature and culture. Though he studied medicine, he did not like clinical practice and quickly turned to studying physiological processes in living systems. Historian of science Philip Pauly's excellent scientific biography of Loeb's work explains Loeb's evolution as a thinker.[9]

Pauly explains Loeb's position in the world of German physiology, physics, and philosophy. Loeb embraced a fully materialistic idea of life but accepted that living systems might have some direction. His regular correspondence with physicists and philosophers imbued in him an engineering perspective regarding biology and the desire to both develop mechanistic explanations in biology and align them with physics. Loeb's search for mechanisms and his penchant for physics were not popular positions in the late nineteenth-century German biological community. Because of his failure to conform to German academic ideas in physiology, and probably also because of restrictions due to his Jewish heritage, Loeb realized that his career prospects in Germany were grim. As he learned more about the American educational system, he decided that he needed to move to the United States. In 1890, he married an American, Anne Leonard, which made the transition all the more attractive.[10]

Despite the fact that the president of Bryn Mawr College, Martha Carey Thomas, was admittedly anti-Semitic, Loeb and his wife persuaded her to hire Loeb in 1891. At Bryn Mawr, he joined Morgan, who taught the morphology-based courses that looked at organismal structure, while Loeb taught physiological courses and apparently very much enjoyed the experience. Loeb remained for only one year, however, reportedly because of the lack of sufficient laboratory facilities. He accepted happily when he was invited in 1892 to move to the new University of Chicago, with its attractive research laboratories and larger group of biological colleagues.[11] Loeb's connection with Morgan and then at Chicago with the biology department director Charles Otis Whitman also took him to Woods Hole and the MBL, where Whitman also served as director.

Loeb spent his first summer at the MBL in 1892, where he organized a series of lectures that became the long-standing

FIGURE 3.5 | Jacques Loeb and his family in Woods Hole, MA. Image courtesy of the Marine Biological Laboratory Archives. https://hdl .handle.net/1912/21129.

Physiology Course (fig. 3.5). That began a tradition that led to his return in most summers even after his move to the University of California in 1902. Realizing that he was very isolated in California, Loeb moved to the Rockefeller Institute in New York in 1910 and continued his summers at the MBL throughout his life, which included helping to run the Rockefeller-funded laboratory at the MBL.

Loeb carried out a number of lines of research, of which we focus on regeneration but also note his work on artificial parthenogenesis. Both Loeb and Morgan carried out research each summer at the MBL. They both looked at a diversity of organisms, asking a range of different questions. Both were immensely curious, and both were opportunistic in the sense of following a productive line of research.

In 1899, Loeb attracted considerable media attention as he published on the ability of sea urchin eggs to develop parthenogenetically, meaning from the egg alone and without having been fertilized. The eggs could develop at least to the larval stage, called the pluteus. He saw this phenomenon as directly revealing how development occurs and as informing processes such as regeneration. In some respects, the egg that develops on its own is replacing a missing part, in this case the sperm cell.

Loeb was especially interested in the effects of external conditions on development and regeneration. Building on studies he and others had carried out in Europe at the Naples Zoological Station, he was curious about the effects on eggs of changing what was called osmotic pressure caused by different salt concentrations. Because changing the salt concentration also changed the internal pressure, Loeb thought that such a change might also affect the speed of development. Sea urchin eggs offered an excellent subject for study since they are relatively large and easy to observe, and they were readily available at the MBL as well as at Naples. Loeb recorded various results, including that when he removed the eggs from concentrated salt water back to sea water, the cells divided quickly, all at once instead of gradually as during normal development. Loeb suggested that for eggs in the concentrated salt solution, the nuclei continued to divide but the reduced water in the cell (because of the increased salt) kept the cytoplasm from dividing. Restoring the cells back to the normal seawater allowed the cytoplasm to respond to the guidance of the nuclei in Loeb's view. He did not regard the cytoplasm as particularly important, whereas the nucleus was and he believed that the mechanical interactions with the nucleus caused development.[12]

In this period around the beginning of the twentieth century, Morgan carried out similar experiments but rejected Loeb's claim that the cytoplasm (which many biologists called protoplasm at the time) was largely irrelevant. For Morgan, the cytoplasm was organized with that system of tensions to which he referred in *Regeneration*. The two argued about these interpretations, along with others working on similar experiments. Loeb happened to be the one to achieve parthenogenesis all the way to the larval stage. As Morgan's student and biographer Alfred Sturtevant reported later, "In later years Morgan sometimes talked about this matter; he clearly felt that Loeb had been secretive about his own work and had used every opportunity to find out just what Morgan was doing. However, Morgan was not as resentful as were some other members of the Woods Hole group on his behalf, and he and Loeb were on close friendly terms during the period when I knew them—from 1913 until Loeb's death in 1924."[13]

Loeb's work on parthenogenesis or, as some news reporters put it, "virgin birth" since development did not seem to require any contribution from the male through fertilization, gained Loeb considerable public notoriety. It is clear that Loeb did not enjoy this attention, which did not result from his own efforts but because others had advertised his work. Yet interest in his work did give him increased leverage for support for his research at Chicago, the MBL, in California, and at the Rockefeller.[14]

Loeb energetically championed a mechanistic approach to

the study of development and continued his exploration of parthenogenesis in different organisms, attempting to explain what factors direct development even without egg fertilization. One factor that particularly struck his interest was tropism. A tropism is the movement of parts or the whole organism in response to an external stimulus. For instance, light can attract organisms to point in a certain direction and to grow in those directions, in what was called heliotropism. Or, Loeb mused, salt concentrations might elicit responses that caused cell division. Loeb wanted to know why. What is going on inside the organism that causes this response to external conditions? His experiments on artificial parthenogenesis were, in effect, a particular example of controlling the external conditions and eliciting an internal response.

In a 1907 article in *Science*, Loeb laid out his emerging ideas. In "On the Chemical Character of the Process of Fertilization and Its Bearing on the Theory of Life Phenomena," he acknowledged that "There may be a difference of opinion as to whether or not it will ever be possible to produce living matter from inanimate; but I think we all agree that we cannot well hope to succeed in making living matter artificially unless we have a clear conception of what living matter is."[15] Loeb wanted to discover the rules governing processes in organisms, then use them to control and improve life. But, as he noted, it was necessary to start by learning how life works. He saw development of both plants and animals as the place to start because, as he noted, here material somehow automatically shapes itself into organized and functioning individual organisms. The most basic question, then, asked what causes this process. And for that, learning how artificial parthenogenesis can produce the same developmental response as normal fertilization would provide key insights.

Since he was a physiologist with a deep interest in physics, it should come as no surprise that most of Loeb's article, as was the case with many others, laid out the chemistry involved. What chemical changes occurred as the individual underwent development? That is, what are the processes within the organism that reflect developmental stages, and also what causes the changes? What guides the formation of membranes, for example? Loeb had an answer. The material in the cell nucleus, or nuclein, contains the physical properties needed to bring about the chemical synthesis of all the necessary parts of the organism. We quote Loeb at some length because this article is very revealing of his approach, as well as how it differed from Morgan's. In Loeb's words,

> I am of the opinion that this mechanism of nuclein synthesis is the thread by which we can find a rational way through the maze of the otherwise bewildering mechanisms, characteristic of living matter; on the one hand, the phenomenon of growth, on the other, those of self-preservation . . . It can be proved that the nucleus itself or one of its constituents acts as a catalyzer in the synthesis of nuclein in the unfertilized egg . . . This influence of the nucleus upon the nuclein synthesis, and the role of this synthesis upon the preservation and continuation of living matter, explains one of the most mystifying characteristics of the latter, namely, the phenomenon of automatic reproduction of cells.[16]

It might seem that Loeb would have been a supporter of the new science of genetics as it developed, but not so. He was intrigued by Mendelism and the mathematics of hereditarian theory but did not approve of what was called cytological genetics, or the emphasis on genetic factors in the nucleus and cytoplasm as determining what happens in the developing organism. By 1916, as Pauly reports based on archival records, Loeb declared that genetics had "begun to bewilder him."[17]

Let's take a moment to reflect on Loeb's way of thinking. From the outset, Loeb emphasizes his search for "characteristics of living matter" rather than "life." For Loeb, life is material; it is matter in motion, just as the materialists we met in the previous chapter believed. His emphasis on living matter is also indicative of his opposition to any form of vitalism. Organisms and their processes, for Loeb, cannot be understood through an appeal to unseen and unknowable forces, as Driesch's vitalistic approach had proposed.

No, life must be material, to be understood through chemistry and physics, and therefore we must adopt a "mechanistic conception of life" in order to truly understand nature, as he emphasized in his book by that name in 1912. In addition to his certainty about materialism, Loeb was also certain about the prime causal importance of the chemical synthesizing action of the nucleus. Loeb revealed his epistemological convictions about how science works with his claim that "it can be proved." Where Morgan offered hypotheses and called for further experimental study, Loeb was busy "proving" that his interpretation was correct. And finally, by doing so, he claimed to be solving that fundamental question of life: what drives cells to divide in ways that seem "automatic"? This 1907 paper laid out the reasoning that underpinned Loeb's continuing work.

In 1912, Loeb published a set of essays as a book under the title *The Mechanistic Conception of Life.* Of the ten chapters looking at various aspects of tropism and other factors shaping development, the first essay lays out his big idea about life. His first sentences offer a bold vision that he continued to develop throughout his career. Again, it is worth quoting him to see the way he words his convictions and purpose. "It is the object of this paper to discuss the question whether our present knowledge gives us any hope that ultimately life, i.e. the sum of all life phenomena, can be unequivocally explained in physico-chemical terms. If on the basis of a serious survey this question can be answered in the affirmative our social and ethical life will have to be put on a scientific basis and our rules of conduct must be brought into harmony with the results of scientific biology."[18]

Yes, these are strong words putting forth a strong view. Loeb meant it. This emphasis on the efficacy of science to provide us with ethical and social guidance as well as biological understanding of life did not persuade many of his colleagues. Yet it did gain him respect. Loeb was asked to direct the Rockefeller-funded physiology lab at the MBL, and when Loeb died, the MBL noted that he became a central figure at the laboratory.[19] He chose to be cremated with his ashes placed in the Woods Hole cemetery at the Church of the Messiah, along with many other scientists of all religions or no religion.

The essays in *Mechanistic Conception* laid out Loeb's values and goals. Another volume a few years later in 1916 asked how

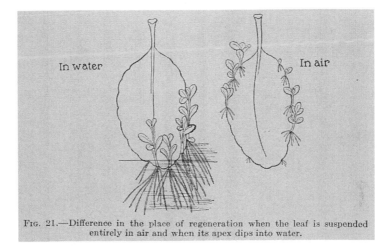

FIG. 21.—Difference in the place of regeneration when the leaf is suspended entirely in air and when its apex dips into water.

FIGURE 3.6 | Figure 21 from Jacques Loeb, *Regeneration*. New York: McGraw-Hill, 1924.

to understand *The Organism as a Whole*. There he put his mechanistic program to work, looking at a range of issues related to life as it exists in the form of organisms, including an extended look at regeneration. He was focused on the materialistic system of the individual organism, on its reaction to damage or injury, and on the ways the responses allowed recovery to become whole and functional.[20] The explanations he sought were deterministic, relying on physical and chemical factors that work to retain the organism as a whole.

Referring to work of Morgan and Child on regeneration, Loeb noted that the ability of an organism to regenerate brings into focus the importance of the whole in shaping parts. His own work had centered on the Bermuda "life plant" *Bryophyllum calycinum* (figs. 3.6 and 3.7). Cutting off a leaf can give rise to other entire plants through processes that do not occur in normal organisms. A leaf does not normally send off roots, for

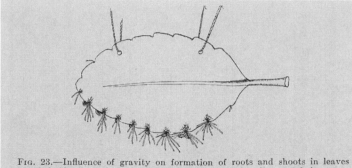

FIG. 23.—Influence of gravity on formation of roots and shoots in leaves suspended sidewise and in a vertical plane in moist air. Roots and shoots are formed only on the lower edge of the leaf.

FIGURE 3.7 | Figure 23 from Jacques Loeb, *Regeneration*. New York: McGraw-Hill, 1924. Images by Miss M. Hedge, Rockefeller Institute Illustration Division.

example, as it does when separated from the rest of the plant. Why not? This was the beginning of years of study of what became a favorite research subject for Loeb. His biographer W. J. V. Osterhout, a colleague at California, reports that Loeb worked on *Bryophyllum calycinum* for years without seeing it growing in the wild, and was delighted when he finally saw it growing in Bermuda.[21]

Continuing his development of his mechanistic conception of life, Loeb also continued his studies of regeneration. His last publication, in 1924 when he died, was *Regeneration*. He opened with a preface reiterating a point he had made repeatedly throughout his career: a successful mechanistic view requires quantitative results, with explanations in terms of mathematical formulas and laws. There is no room in the biological sciences for vitalism or any other metaphysical speculations such as "guiding principles" like those that Driesch offered. Nor is there room for the kinds of hypotheses like

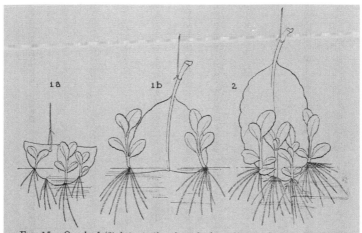

FIG. 15.—One leaf (2) intact, the sister leaf cut into 2 pieces, a small apical one (1*a*), and a larger basal one (1*b*). Shoot and root production are in proportion to the mass of the pieces. Mar. 26 to Apr. 17, 1923.

FIGURE 3.8 | Figure 15 from Jacques Loeb, *Regeneration*. New York: McGraw-Hill, 1924.

that of postgeneration that Roux offered. And of the admittedly considerable work done to date, Loeb noted that none had been quantitative. The point of this volume was to bring together the results of his quantitative experiments carried out over the years.

In particular, Loeb focused on the weight of plants under different conditions. He held that there is a defined quantity of mass in a plant. A stem or a leaf has a particular quantity of material available to promote growth, he felt, and if part is cut off, then the remainder has sufficient mass to carry out growth and thereby to replace the missing part (fig. 3.8). "The process of regeneration was thus revealed as a purely physico-chemical phenomenon, leaving no necessity nor room for the postulation of a guiding principle aside from the purely physico-chemical forces."[22] The "mass relation" was the key for interpreting the regenerative processes.

Loeb addressed two themes with two separate parts of his book. First came injury, or as he put it mutilation, and regeneration based on the mass relation. Second was the idea of polarity across the organism, a theme to which we will return with Child, who focused on what he called gradients. Loeb affirmed that experiments carried out throughout his career provide a "proof" of his theory.

We need not go through the reasoning in detail, but it is worth trying to understand what he meant by the mass relation on which he depended heavily. After a brief review of earlier work, including Morgan's planarian experiments, Loeb stated definitively that no scientific explanation had emerged for regeneration, "if by a scientific explanation is meant a rationalistic mathematical theory based on quantitative measurements."[23] Previous attempts at explanation were just words.

In order to provide the mathematical explanation that he required, Loeb started with three assumptions. First, light causes the stems and leaves of his *Bryophyllum calycinum* to have all it needs to grow in the right way, with all the requisite substances. Second, the mass of the plant material increases in proportion to the amount of chlorophyll. Third, the amount of chlorophyll available for the plant to regenerate remains constant throughout the process. These assumptions give us what Loeb called the "mass relation," according to which "the fact that the mass of shoots and roots regenerated varies in proportion with the mass of the leaf or stem where the regeneration occurs."[24] Where regeneration occurs depends on the placement of "anlagen," but how it occurs depends on the mass. For Loeb, this mass relation functions as a law, and he felt it was sufficiently simple and clear to explain regeneration in a way that should be compelling. The fact that not everybody, and indeed not many, developmental biologists took up his reasoning suggests that Loeb was overly optimistic in his assessment.

After working through the series of experiments and their interpretations, Loeb concluded that his approach explains regeneration while leaving two questions for additional research. The mass relation does not explain where the regeneration occurs — that is, why it occurs only at particular points on the leaf. Also, he acknowledged that the mass relation does not explain why the parts that regenerate are always precisely those of *Bryophyllum calycinum* and not some other kind of plants. These points, he maintained, could also be addressed in physico-chemical terms but required additional studies. Unfortunately, as a result of becoming ill while vacationing in Bermuda where those *Bryophyllum* grow in the wild, he died in 1924 and his studies ended. As he would have wanted, he was actively and energetically carrying out research until the very end. His mechanistic emphasis drew considerable attention, as did his work on artificial parthenogenesis. His ideas about the mass relation and regeneration had less impact and are certainly not recognized in textbooks in the way that Morgan's work on genetics has been. Yet Loeb's passion for understanding and controlling life does resonate with current enthusiasm for efforts to synthesize life in the laboratory through the field called synthetic biology.

## CHARLES MANNING CHILD

Our third figure for closer consideration is Charles Manning Child. Whereas Morgan grew up in Kentucky wandering the countryside and becoming fascinated with nature, and Loeb grew up in Germany immersed in philosophy and a culture that he eventually found unwelcoming, Child had an apparently happy childhood in New England. Child's research intersected with both Morgan and Loeb frequently and in various ways, and yet he adopted a different approach just as he had had a different path to biology.

Although Child was born in Ypsilanti, Michigan, in 1869, he grew up in Higganum, Connecticut, as essentially an only child since his siblings all died in childhood. In high school he became interested in natural history and explorations made possible by using the microscope. His parents died, and after earning a bachelor's and a master's degree from Wesleyan College, he went to Germany from 1892 to 1894 to learn about science as well as culture. There he received a PhD from the University of Leipzig, working with zoologist Rudolf Leuckart. All of this shaped his outlook on what science could be, and the experience put him in a good position to enter the newly emerging profession of biology in the United States. Where Morgan had earned his credentials at the Johns Hopkins University, and Loeb in Germany, Child built on his German degree at the University of Chicago.

The Johns Hopkins University was the first American university, founded in 1876 on the British model of a research university, with the biological sciences centrally placed in its graduate programs. The University of Chicago was founded in 1890, building on German models of education that emphasized research, and also with the biological sciences as central. Whereas Johns Hopkins emphasized life sciences alongside their medical school, the University of Chicago was envisioned as a school for liberal arts and sciences and did not have an affiliated medical school. John Boyer's *The University of Chicago: A History* details the vision and values for what quickly became a premier American university.[25]

At Chicago, President William Rainey Harper recruited Charles Otis Whitman as professor of zoology to establish the biology program. Whitman, as it turns out, had received his PhD from the University of Leipzig, studying under Rudolf Leuckart just as Child did two decades later. Child joined the University of Chicago in 1895 with titles that advanced from zoology assistant to associate to instructor to assistant professor in 1905 and on to associate professor and then full professor in 1916. Child remained at Chicago until he retired in 1934. He obviously intersected there with Loeb, who was at Chicago from the time Child arrived until Loeb left for California in 1902.[26]

In addition to the lively research community and commitment to excellent education at Chicago, Whitman enticed those who had any interest in embryology to join him for the summer at the MBL. As MBL director, Whitman invited his faculty members to attend as investigators and course instructors, and he expected his students to take the courses and explore research possibilities. Whitman also encouraged everyone to study embryology by starting with cell lineage studies, which involved taking eggs, fertilizing them, and watching as they developed through each cell division. With the right kinds of eggs visible through the microscope, the observer could follow the lineage of each cell. Whitman saw that by collecting observations from a wide range of organisms, the community could compare the patterns of division and begin to document differentiation and development. Whitman himself had studied leeches, Morgan studied the flatworm *Polycoerus* though he never wrote up the results, and Child looked at the annelid *Arenicola*.[27]

At the MBL following his first year at Chicago, Child also served as an instructor for the Embryology Course. While there, Child met Lydia Van Meter, who was one of the students in the course; they married in 1899 and had an apparently happy life together. Child returned to the MBL for several years, but he reported that he found the available research material more favorable on the West Coast. Archival letters among a number of MBL scientists suggest that Child never quite "fit in." Perhaps this happened in part because of his different ideas or perhaps for other personal reasons. He never bought a house in Woods Hole, and he did not take his family and a cohort of students year after year to the MBL as Morgan, Loeb, and so many other American biological leaders did. Perhaps he preferred to stay home in Chicago and write. He had quite a productive career, and on regeneration, he produced roughly forty papers in the period from 1900 to 1910. These looked especially at development in coelenterate flatworms, namely, the planarians that Morgan and Loeb had also studied. Many of these appeared in Roux's edited journal or in the journal edited at the MBL, the *Biological Bulletin*.

During his teaching at the MBL and Chicago, and with his own research, Child became committed to understanding how development works. What drives development such that the parts become organized functioning wholes of just the right sort? This led him to become immersed in the various theories of materialism and vitalism, as well as questions about the extent to which each organism is self-organizing because of internal predetermination of some sort or more driven by responses to external environmental factors. Like Loeb, he wanted to understand the interactions within the organism, with questions about what makes each one work as an individual. Morgan had pointed to his idea of "tensions," Loeb had his tropisms and mass relations, and Child, as we will see, had his gradients.

In November of 1915, Child published *Individuality in Organisms*. Earlier that year, he had published his much longer *Senescence and Rejuvenescence*, a densely detailed presentation of the life cycle of an organic individual. Fortunately for those who seek to understand what Child considered most important in his work, the University of Chicago Press initiated its "Science Series," for which each volume would "present the subject in as summary a manner and with as little technical details as is consistent with sound method. They will be written not only for the specialist but for the educated layman." Child clearly had to resist his usual tendency to include every detail, and he focused his readable short book on "the nature of the unity and order in the organism, the constancy of character and course of development, the maintenance of individuality in a changing environment, and the processes of physiological isolation, disintegration, and integration in reproduction."[28]

There, Child started with the observation that life consists of individual units. Cells can be individual units, and organisms are also individual units. Individuality consists of being alive, being of limited size, having a definite form, and being dynamically coordinated. The result is a unity of the whole out of parts, and therefore we need to understand the nature of the unity of the whole. Child pointed to polarity and symmetry as helping to define the unity within the organism as it develops through successive stages: polarity from end to end, symmetry from side to side or around a central point. The problem for biology, then, was to recognize that individual units come together into a cohesive, functioning whole, such as cells within an organism and then to ask how.

To address this question, he pointed briefly to theories of organic individuality advanced by others before proceeding to his own interpretations. First, he pointed to Weismann's focus on the germ plasm and heredity. Yet, as Child noted, interpretations such as Weismann's and Roux's all rely on speculative hereditary units of some sort and therefore do not produce a scientific and testable theory. Rather, "They merely translate the problem into hypothetical terms which are beyond the reach of scientific method."[29] Second, he looked at vitalistic theories that invoke some kind of nonmaterial principle that controls the processes. He pointed to Driesch as the primary proponent still holding on to a version of this view that had been largely discarded by almost all other biologists. Such ideas also remain necessarily speculative and are not satisfying as scientific accounts either, Child explained. A third option was the physico-chemical approach to explaining life processes in terms of matter and motion, such as Loeb so enthusiastically maintained. Though appealing in being scientific at least, Child felt that such accounts had not yet yielded the laws said to be the goal of the approach. The organism is far more complex than a crystal or other simple material system that the physico-chemical approaches tended to suggest, and the approach therefore also failed in not being able to provide an explanation for the very real changes that occur. Therefore, these attempts all fail.

Perhaps the unity of individual lives does not really exist, as some critics had suggested, but Child rejected that idea by pointing to the way that organisms behave in what must be coordinated ways. None of the existing hypotheses could explain the interactive dynamic nature of life, which suggests that there is something transmitted within the organism, a kind of stimulus. Child held that the organism has regions with high metabolic rates and others with lower rates, and some form of transmission carries material from the higher to lower regions. According to this view, some stimulus, perhaps from outside the organism, initiates the transmission. This is a transmission of "excitations" rather than of some substance in particular, Child suggested. As a result, a gradient becomes established throughout the organism. He called this gradation of greater or lesser intensity across different factors a metabolic gradient—a term that biologists have accepted since. He explained these gradients and their importance in the rest of the book.

Child's idea of gradients within organisms provided an interpretation that paid attention to the unity of the whole individual organism and provided an account of differences that exist throughout the individual. In some ways, the idea was responding to the same impulses as Morgan's suggested system of tensions. Yet Child's ideas are more systemic in transferring causal actions throughout the entire organism, and the ideas are more dynamic in reflecting and causing change over time. He explored gradients within organisms through a variety of means. In one experimental setup, Child measured the response of a diverse array of organisms to the presence of cyanide. He reasoned that areas of high metabolic activity would acclimate more readily to the poison, whereas areas of low metabolic activity experience the inverse. By observing the state of the different cells and organs throughout the body under the poisonous conditions, he provided evidence to support his claim. Child also showed that the same principle of susceptibility gradients was at work within embryos (fig. 3.9). Using various chemicals, he was able to inhibit the growth and development of different parts of developing organisms.

Child did not use the term regeneration. He did, however, consider the phenomenon extensively within his exploration of gradients and used the term "reconstitution" instead. Like Morgan, Child came to the conclusion that regeneration was the same process as development: "The reconstitution of pieces into new individuals is fundamentally the same process

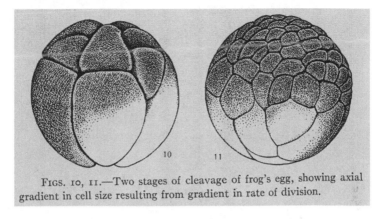

FIGS. 10, 11.—Two stages of cleavage of frog's egg, showing axial gradient in cell size resulting from gradient in rate of division.

FIGURE 3.9 | Figures 10 and 11 from Charles Manning Child, *Individuality in Organisms*. Chicago: University of Chicago Press, 1915.

as embryonic development, and the same relation of dominance and subordination exists in both."[30] Though his small volume included a look at various different kinds of gradients and several different kinds of organisms, Child included what became a favorite, planarians (fig. 3.10). Morgan had chopped heads and tails and other pieces off these flatworms, and so had Loeb. Child apparently found these creatures just as fascinating. Chop off the head, and normally a head grows back. Chop off a tail, and a tail emerges.

Just observing the patterns of response to injury seemed to Child to demonstrate the existence of gradients. He concluded that the evidence showed clearly that "axial gradients in the dynamic processes are characteristic features of organisms and that a definite relation exists of organisms and that a definite relation exists in each individual between the direction of the gradient of any axis and the physiological and structural order which arises along that axis."[31] Proceeding to look at the ways

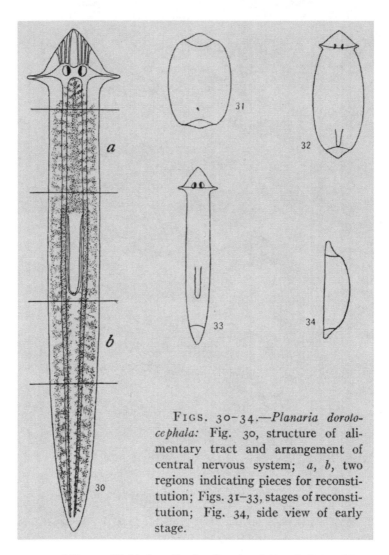

FIGS. 30-34.—*Planaria doroto-cephala:* Fig. 30, structure of alimentary tract and arrangement of central nervous system; *a, b,* two regions indicating pieces for reconstitution; Figs. 31-33, stages of reconstitution; Fig. 34, side view of early stage.

FIGURE 3.10 | Figures 30-34 from Charles Manning Child, *Individuality in Organisms*. Chicago: University of Chicago Press, 1915.

that gradients work, with some dominant and others subordinate, Child sought to lay out how morphogenetic development works. That is, how does the organism form from an initially unstructured (or at least not obviously structured) egg cell to a highly structured individual of just the right sort? He sought to understand the nature of dominance and the extent of the dominance with numerous experiments. His goal, he proclaimed, was to integrate the understanding of structure that had dominated embryological studies with an understanding of the dynamics of change that had largely been set off as a matter for physiology. Organisms have structure and function, and it was time to bring them together to understand changing factors that shape and preserve the individuality of organisms. This small book laid out his ideas, which he continued to develop and which appeared in more detail in his much later volume *Patterns and Problems of Development* in 1941.

## REGENERATION TO MID-TWENTIETH CENTURY

Among those studying regeneration in the first couple of decades of the twentieth century, we have focused on three Americans who show a community of researchers interacting and exchanging ideas. They were all central to the debates about what life is, asking what an organized unit of life is and how it develops. As a complex system, what keeps an individual organism and its parts whole and functioning, despite disruptions through injury, damage, or failure in some way? Regeneration was a lively topic into the 1920s, then it faded from its central position in developmental biology. Yet the focus on systems-based thinking that Morgan, Loeb, and Child had each embraced in different ways remained.

These three envisioned organisms as systems of interacting parts and regeneration as a process accomplished through some sort of regulation by that system. These scientists wanted others to test their ideas about how regeneration works (Morgan, Child), or turn biology into a quantitative science (Loeb). By the 1920s, they had clearly laid out the idea of an organized individual organism that reacts to stimuli internally and in response to the environment. That some form of tensions (Morgan), tropisms (Loeb), and gradients (Child) must be involved had also become clearly articulated and largely accepted by developmental biology, neurobiology, ecology, and other biological fields.

But, in the decades following Morgan's, Loeb's, and Child's proposals, the scientific community failed to follow through on testing these proposed explanations or generating new ones, and developmental biologists incorporated very little quantitative analysis. As developmental biologist Lewis Wolpert acknowledged decades later in 1991, despite Morgan's, Loeb's, and Child's excellent earlier contributions, "by the 1950s the understanding of gradients had made little progress. Few clearly worked out models or mechanisms existed, and there was a notable lack of quantitative analysis."[32]

Instead, those researchers who continued to study regeneration of individual organisms tended also to follow in the tradition of focusing on morphology and description. They looked for the particular locus of the cells capable of regenerating. Biologist Richard Liversage reflected that studies of regeneration in amphibians such as salamanders (urodeles) typically pointed to the same idea as Bonnet or Spallanzani had sug gested so much earlier, namely, that regeneration is guided by some kind of "formative stuffs." By the mid-twentieth century, they called those stuffs by the more scientific-sounding name of the blastema to identify that particularly productive area of the organism capable of generating cells and regenerating damaged cells.[33] Developmental biologists acted as if they just needed to identify these particularly active areas, then observe and analyze them more and more, and somehow they would reach explanations of regeneration. Wolpert did not agree with that limited approach, nor did a growing group of others.

This very morphological approach characterized most study of regeneration of organisms, including a new flurry of interest caused by the hope of finding ways to heal injured soldiers in World War II. If only researchers could discover what allows wound repair in an injured organism, they might apply that information to repair human wounds. Yale embryologist Ross Granville Harrison tried with the last paper he published in 1947. He removed part of the neural plate and observed the movement of cells to repair the wound: where did those cells come from, and were they already existing cells just moving around or newly derived cells? Even though he drew on wax models to try to observe the internal workings, his approach remained morphological and descriptive.[34] The hope for applications remained high, but reality lagged behind aspirations.

As Wolpert noted, and as Loeb would have regretted, these researchers failed to carry out quantitative measures of anything. They did not develop models of what they saw in ways that could identify patterns of movement or generate predictions. They did not seek causes of regeneration beyond identifying the cells involved or asking for a description of what initiates and guides the regenerative process. They did not provide mechanistic explanations for why some cells, tissues, and organisms can regenerate while others cannot. The approach to regeneration of individual organisms only changed after biology incorporated genetic and evolutionary approaches, with computational approaches, models, and much clearer ideas of the systems involved in development and regeneration. It took many decades, to the end of the twentieth century, before researchers returned to earlier questions using new experimental and interpretative methods.

In 2020 in *The Dance of Life*, University of Cambridge developmental biologist Magdalena Zernicka-Goetz and British science writer Roger Highfield emphasized the ongoing studies of early developmental stages in a single embryo. The work retains a focus on individual organisms but now looks at cells much more explicitly as living systems that come together to produce embryos as complex living systems. These authors point to decades of assumptions about how the cells self-organize to become embryos and show how current research in the laboratory and with models and comparisons demands changing assumptions.

Intriguingly, Zernicka-Goetz's work shows new and diverse ways that embryos can undergo damage or loss and nonetheless regenerate. As she describes the complex dance of life, she sounds rather like Loeb: "the dance can now only be observed but not manipulated, and it is most easily visualized in terms of collective migrations of cells."[35] And she seems to feel so much closer to being able to carry out that manipulation than Loeb was.

Zernicka-Goetz's approach involves signaling that draws on factors like Morgan's tensions or Child's gradients. Physico-chemical signals within the organism interact with particular cells to stimulate them to respond and start changing. This developmental process leads to differences among the cells and therefore an organization of the cellular parts into the whole organism. Just as Morgan, Loeb, and Child felt, development must involve self-organization, driven by factors internal to the organism itself. She shows the ways that the parts of the developing organism interact as a complex system to make regeneration possible and effective.

Further probing of the complex systems of cells, signaling, and positional effects occurs in work by Harvard biologist Mansi Srivastava and others. Drawing on extensive comparisons and probing of evolutionary adaptations has allowed them to interpret "the regulatory landscape of whole-body regeneration" in the order of small invertebrates called the acoels.[36] Connecting development with evolution also raises questions about environmental factors influencing past adaptations and brings us to another view of intersecting scales of living systems. Study of regeneration more generally across all scales of living systems increased significantly as the twentieth century progressed and gave way to the twentieth-first century.

# 4        Living Systems and Different Scales

All living systems maintain some capacity to respond to injury and heal themselves in the face of damage. Understanding how the different systems do this is crucial to our continued existence on this planet and requires understanding how regeneration operates and what parts, relationships, and logic(s) are necessary in order for regeneration to take place properly. Every day people sustain debilitating injuries from accidents or violence, or they suffer degeneration at the hands of illness. Every day our planet sustains more damage to its ecosystems. As Jacques Loeb recognized, we will benefit tremendously from ways to reengineer ourselves and our environment and begin to repair the damage.

Scientists are already at work and have been carrying out research to understand and use regeneration to heal injured living systems for a long time, as we have seen. But, and this is crucial, we have lacked a perspective of regeneration as a phenomenon of *all* living systems. We have lacked an understanding of a logic of regeneration that makes what we have learned at one scale, like salamander tails or neurons, transferable and translatable to other scales of living systems, like ecosystems. At each scale, regeneration follows a set of rules that we saw Morgan, Loeb, and Child seeking as recounted in the previous chapter. The larger question is how best to discover what these sets of rules at individual scales have in common, that is to find a shared logic of regeneration. Without this kind of knowledge and transferability, scientists will continue their research and make some advances, but we will not be able to make the kinds of breakthroughs that our broken bodies and ecosystems need right now. So how has our thinking gotten from the emerging systems-based regeneration ideas of Morgan, Loeb, and Child to the all-encompassing and cross-cutting living systems-based approach to regeneration that we are advocating today?

In the previous chapters, we have seen how thinking about regeneration evolved over time. We moved from the views of Aristotle, for whom regeneration was part of a larger worldview of constantly changing matter, to the Enlightenment, where regeneration was tied to understanding generation and to debates about vitalism versus materialism and epigenesis versus preformation. Regeneration at the start of the nineteenth century became bound to experimentalism and was considered a phenomenon in and of itself though still as a continuation of generative processes. Scientists like Morgan, Loeb, and Child began to envision organisms as systems and saw each system's parts as interacting in order to adapt to changing environments in some way. Yet deeper understanding of the general rules governing how the parts interact within the systems remained elusive.

The situation slowly began to change after World War II

and began to reach fruition by the end of the twentieth century. A convergence of genetics based in DNA, evolution within the movement called the "evolutionary synthesis" because it brought together many previously separate ideas, and expanded ideas about how life might best be understood based on systems-based thinking all began to encourage new approaches across the different specialized fields of biology. Various definitions of systems arose, including some focused on organized collections of information and even related to mathematical modeling of parts that make up systems. These different ways of thinking about systems have been advanced at different times and in different contexts during the twentieth century and even led to an approach called systems biology with an emphasis on modeling and quantitative analysis of systems and their components. Our approach in this book has remained historical, focused on how biologists understood biological units over time, and has advocated understanding systems as a group of parts that interact in a coordinated fashion. In biology and medicine, the earliest uses referred to organ and functional systems such as the nervous system or reproductive system.

What changed by the middle of the twentieth century is how researchers studied biological systems. Geneticist François Jacob noted in 1974 that "Every object that biology studies is a system of systems." Plant biologist Anthony Trewavas pointed to Jacob's observation and to other earlier ideas of organic systems thinking that had not developed as fully as he thought they should have.[1] What plant biology needed, he urged, was more robust computational modeling of the complex systems, and computational modeling was dependent on understanding or hypothesizing the nature of the relationships among a system's parts.

Yet even in 2020, the US National Institutes of Health acknowledged that this admirable goal is not so simple to achieve because it involves many different steps. At the NIH, National Institute of Allergy and Infectious Diseases Director Anthony Fauci and Scientific Director Kathy Zoon dedicated resources to "embrace experimental and computational techniques to explore connections in all their intricate glory." In describing this NIH approach to understanding immunological systems in particular, author Christopher Wanjek says to "start with computational modeling." This requires bringing together information from different sources, compiling it, making it compatible, then computing things. Computing what? For example, computing the steps in the process by which cells communicate with each other, called cell signaling pathways. Then add "One Generous Serving of Proteomics" because it's important to add information about what protein products are emerging from cell action. "Mix Well With Genomics," and "Oh, Right, Immunology, Too." Then "All Together Now," the resulting models can understand and predict responses to the complex process of vaccination, pulling together intersections of gene networks and adaptive responses. That's the NIH goal: get at all the different parts and processes of a complex living system. Work proceeds and fully articulated models have not yet emerged and persuaded everybody to adopt them, but here we find a very clear explanation of a systems approach to a complex problem that requires pulling together different kinds of information, generated with different approaches, and all part of understanding the complex adaptive system.[2] A systems-based approach to regeneration requires the same approach of sorting out all the parts and then putting them together.

As we have described, historical understanding of regeneration began at the scale of the organism. The system can experience injury or failure, and it responds. In some cases, it responds by regenerating the affected part. In other cases, a wound fails to heal, a lost part fails to regenerate, or the organism may even die. The interacting whole seems to direct regulation of the parts, not quite in the special vitalistic ways Driesch suggested in his interpretation of regulation, but recognizing the interactions among the connected parts of the whole in ways that Driesch was trying to explain.

In the 1950s and 1960s, scientists such as Jacques Monod, François Jacob, Eric Davidson, and Roy John Britten began to create logic models for how biological systems work in the language of genetics. These scientists offered models that could explain and predict and that understood organisms as systems of interacting parts. In one model that they called the lac-operon model, Monod and Jacob explained how single-celled organisms called prokaryotes processed the sugar lactose. In another, Davidson and Britten drew on sea urchin development to model the growth and differentiation of multicellular organisms during development and predicted the kinds of genes required.[3] By the end of the twentieth century, systems biological approaches involving modeling and articulation of the rules and logic of life in computational terms had become foundational for science. A demonstration of this came in repeated discussions when our colleague Manfred Laubichler and Jane Maienschein were sitting on Davidson's outdoor terrace in Pasadena, California. Davidson teased Maienschein that the approaches of those "old embryologists," such as Morgan, Child, and even Loeb, were "just phenomenological" because they had failed to generate models and therefore failed to offer either explanation or predictive power. Nor, he pointed out, could they have achieved control with the information they had, as Loeb so eagerly desired. Davidson's and Britten's concept of modeling regulation during development is crucial to envisioning regeneration as a systems-level regulatory process, and Davidson spent the decades after introducing the model in 1969 (fig. 4.1) until his death in 2015 accumulating the data to fully tune and test his model.

Beyond gene regulatory networks, developmental biologists in recent decades have also recognized the importance of epigenetic networks and signaling across cells. Epigenetics at its simplest follows Aristotle's emphasis on the gradual unfolding of development over time as the individual organism follows its internal causes and also responds to the environment. Epigenetics as a field of biology emerged in the early 2000s and refers to heritable changes that are not based on underlying changes in the DNA. Much research continues to elaborate the ways in which inheritance can work beyond DNA, and the work has reminded those studying development to pay attention to all parts of the cell, the ways that cells interact, and the nature of the resulting whole. We have a developing system with parts that include DNA, cells, and cell interactions. The

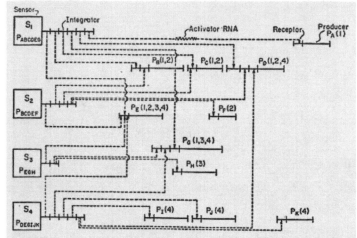

Fig. 2. This diagram is intended to suggest the existence of overlapping batteries of genes and to show how, according to the model, control of their transcription might occur. The dotted lines symbolize the diffusion of activator RNA from its sites of synthesis, the integrator genes, to the receptor genes. The numbers in parentheses show which sensor genes control the transcription of the producer genes. At each sensor the battery of producer genes activated by that sensor is listed. In reality many batteries will be much larger than those shown and some genes will be part of hundreds of batteries.

FIGURE 4.1 | Figure 2 from Roy J. Britten and Eric H. Davidson, "Gene Regulation for Higher Cells: A Theory," *Science* 165 (1969): 349–57. Reprinted with permission from AAAS.

system follows rules that go beyond genetics and that researchers are still working out. This is an exciting innovative approach to development, and one with which Morgan, Loeb, and Child would have been quite comfortable.

Planarians were the workhorse for the regenerative studies of Morgan, Loeb, and Child but were then largely set aside by most others during the following decades. Yet new ways of approaching developmental questions have brought these flatworms back to center stage as planarians offer valuable insight into regulatory processes and systems-level understanding of development and regeneration. A leader in this work, Alejandro Sánchez Alvarado at Stowers Institute for Medical Research, gives us a great example of building on historical work to carry out first-rate science today.[4] Sánchez Alvarado is explicit about his debt to predecessors, including Morgan, Loeb, and Child. But whereas none of them examined genetic bases for development and regeneration, and none included evolutionary thinking, Sánchez Alvarado adds both through studies called "Evo Devo."

Drawing on advanced tools for making observations and collecting data, as Morgan, Loeb, and Child did, Sánchez Alvarado adds experimental manipulations of cells and makes microscopic images of how they respond to the changes. He recognizes that those cells that allow regeneration to occur are specialized cells called stem cells, which are undifferentiated and respond to the needs of the system. His systems consist of genes, cells, signals from genes within an organism, and signals among cells as well as responses to the environment of each cell and of the organism as a whole. With his approach, his extensive study of these by-now familiar planarians provides basic scientific knowledge about development and how it responds to damage. Because "we all have a bit of 'planaria' hidden in our genomes," as he puts it, knowledge about how these flatworms regulate those particular genes also informs what happens in humans. This, in turn, suggests ways to apply the knowledge for human medicine or to engineer and control life, as Loeb envisioned (https://www.stowers.org/faculty /s%C3%A1nchez-lab). As Sánchez Alvarado notes, "To solve old problems, study new research organisms," meaning that understanding a phenomenon like regeneration requires looking broadly at how it manifests and works across living systems.[5]

Looking across living systems for access to the fundamentals of regeneration need not be limited to examining the process across organisms but can accommodate larger scales. To gain further understanding of why some parts, but not others, regenerate and to explain in more detail how they carry out regeneration, researchers have also looked more closely at smaller systems within the whole organism: stem cells, the nervous system, and germ line cells, for example.

In 1998, two lines of research led to what was called human embryonic stem cell research. University of Wisconsin researcher James Thompson reported the first successful isolation and culture of stem cells in human embryos, while John Gearhart at the Johns Hopkins University announced similar results from germ line cells in human fetuses. They explained the possibilities for being able to manipulate stem cells to become medically useful different kinds of cells such as heart muscle, pancreatic, nerve, or others. This suggested a tremendously exciting type of regeneration through controlling development that Loeb wanted.

Yet the term stem cell and its potential for regeneration has been used in essentially its current meaning since the 1950s. At first, researchers realized that certain cells, labeled hematopoietic cells, were not yet differentiated but were capable of becoming blood cells. The idea that some cells that remain in the body, or are even generated later in life, are not differentiated as specific kinds of cells raised questions about how often that occurs and when and how and why. Developmental biologists through the 1960s and 1970s looked at mouse embryos to identify which cells remained undifferentiated and just how much flexibility they retained. Yet the mechanisms that controlled or regulated the differentiation process remained elusive.[6]

Through the last half of the twentieth century, a taxonomy arose of kinds of stem cells with different degrees of differentiation. Totipotent stem cells are those with the capacity to become a whole organism under the right conditions and regulated in the right way. In his 1891 experiments, Driesch had referred to his two sea urchin cells that each became a small whole pluteus larva as totipotent. Pluripotent stem cells have the capacity to become any kind of cell, but not all. In other words, they remain very flexible but any one pluripotent stem cell cannot become a whole organism. Whether they do become, say, a heart or a nerve cell depends on the environment in which they exist and the factors acting on or within them. Multipotent cells are more limited still, with the ability to become any of several different kinds of cells, but not just any kind; they might have the capacity to become one kind or another kind of nerve cell, but not a heart cell or kidney cell, for example. And unipotent cells are just like they sound; they are not differentiated yet, but when the conditions are right, they can become just one particular kind of cell. All these different kinds of stem cells show that each cell has a particular capacity itself, and it realizes its potential to become a particular kind of cell in response to the conditions around it. Each cell is part of a larger networked system that helps regulate and control what happens to each part.

When news about the first cultivation of human embryonic stem cells burst on the scene in 1998, there was great excitement about the possibilities because it suggested that finally researchers were close to achieving Loeb's goal of engineering life. Some who objected to using stem cells that came from embryos and therefore necessarily involved killing the embryos were horrified. There were also enormous debates around policy, practice, and personal choice. Writing in *The New York Times*, reporter Nicholas Wade introduced the science and pointed to resulting ethical questions.[7] Since the cells could be cultivated in the lab, in a glass dish, then feeding different food or culture media should produce different kinds of cells since we really are (at least in part) what we eat. With a bit more knowledge, scientists should be able to produce any type of cell out of these embryonic pluripotent undifferentiated cells.

California developed a major research initiative to discover ways to implement regenerative medicine, the California Institute for Regenerative Medicine (or CIRM). Before 1998, the term "regenerative medicine" would not have meant anything in particular but just pointed to possibilities and an assortment of small efforts to control regeneration as with the hematopoietic stem cells in the bone marrow for treating leukemia. After 1998, Regenerative Medicine gained a life of its own. While the earliest efforts focused on using stem cells from embryos, gradually political and practice pressures pushed researchers also to use adult stem cells. In fact, they discovered that adult stem cells are far more numerous and more diverse than had been thought. In addition, it turned out to be possible to reprogram adult cells to become stem cells through regulating genes, for example, by adding new genes that could produce different proteins.

One particularly powerful example of how stem cells have been used in regenerative medicine is the case of X-linked severe combined immunodeficiency (X-SCID). X-SCID is an inherited disorder of the immune system that occurs almost exclusively in males. X-SCID is caused by a mutation in a gene, which under normal circumstances makes a protein that is necessary for the growth and maturation of the immune system. Children with X-SCID are prone to recurrent infections and, without treatment, do not usually live past infancy. Since 1990, scientists have sought to use gene therapies to cure this disease. In 2000, a group of researchers in France reported the first successful attempt to edit the malfunctioning gene responsible for this disease.[8] The procedure included harvesting the hematopoietic stem cells from affected children, editing the malfunctioning gene, and then reintroducing the edited stem cells back into the patients. After a period of time, the edited stem cells did their job and regenerated a healthy functioning immune system in the affected children. The NIH program called the Regenerative Medicine Innovation Project offers an overview to current and evolving research in this area (see https://www .nih.gov/rmi, accessed August 11, 2020).

The gene editing treatment of these hematopoietic stem cells was hailed as a "landmark" and a "proof of principle" for gene therapy and for using stem cells within regenerative medicine. However, just two years after the therapy successfully restored their immune systems, two of the recipients of the treatments developed leukemia caused by the gene therapy.[9] The exact mechanisms for how the gene therapy caused leukemia remain unknown, but the nature of stem cells themselves was implicated. This example should give us pause to reflect on our understanding of living systems, like stem cells, and how each part of a system interacts with other parts during regenerative processes.

Looking even further back than the discovery of stem cells, scientists have been working on understanding regeneration within the nervous system since the turn of the twentieth century, although research in this area really took off during World War I. In 1928, for example, one of the earliest proponents of neuron theory and observers of nerve cell regeneration, Santiago Ramón y Cajal, wrote *Degeneration and Regeneration of the Nervous System*. His work was close to the approaches taken by Morgan, Loeb, and Child, with a focus on individual organisms and the functioning set of networked cells in that individual. The idea of the system was similar, and a system could remain functional or could suffer injury and fail. But Ramón y Cajal and his contemporaries did not yet have genetic information, evolutionary accounts, or computational tools that leading researchers use today in regenerative medicine.[10]

Research on regeneration in the nervous system continued throughout the twentieth century, with little progress toward clinical applications. The prospect of using embryonic stem cells was thus immediately exciting to anybody with a spinal cord injury, diseases in which nerve cells had degenerated such as Parkinson's disease, or in other cases of nerve loss such as Alzheimer's. "Superman" Christopher Reeve, who had suffered a severe spinal cord injury, and the actor Michael J. Fox, who was diagnosed with Parkinson's disease, both became spokesmen and advocates for research that might bring treatments for so many people.

In the context of excitement about stem cell research and applications, biologists have begun to look more closely, in different ways, and not just at what we might do with stem cells for medical applications. They ask also what we can learn from the natural processes of regeneration, such as in the jawless fish lampreys or other animals. What can those animals do when they are injured or damaged to restore their functioning system? What can we cause them to do by adding or manipulating cells? What are the sets of rules, or logic, governing regeneration?

Our next example concerns germ line cells, which are the reproductive cells within the body (sperm and ova) that allow sexually reproducing species to have children. These cells have been a major focus of ideas about correcting genetic problems through regenerative medicine and for ideas about engineering life. Normally, they remain inside individual humans, and unless they are used quickly, the cells die when they are taken out for experimental or clinical purposes. This is a problem for patients undergoing radiation treatments for cancer, since the radiation can damage or kill the germ line cells and keep the patient from being able to reproduce them in the future. For decades, freezing the cells, or cryopreservation, has allowed those patients to store their germ line cells for future use through assisted reproduction technologies. Recently, scientists have begun to cultivate other types of cells, such as skin cells, from such patients, with the hope of reengineering them into viable germ line cells that could also be used with assisted reproduction technologies or even full regeneration of reproductive capabilities in patients whose germ line cells have been damaged.[11] This type of experimental creation of germ line cells has been highly successful in mice and has recently been extended to human studies carried out both in the lab *in vitro* and in living patients *in vivo,* asking how the female ovaries and male testes might be enticed to regenerate germ line cells after damage.[12]

Since the late nineteenth-century work of August Weismann, the germ line cells have been assumed to be separate from the rest of the body's cells, also called somatic cells, and insulated from effects of the surrounding environment. Weismann was so persuasive that most biologists today repeat without much thought his assumptions that the germ line cells and somatic cells are separate and distinct and that somatic cells cannot become a part of the germ line once it has formed.[13] This has led to the conviction that germ line cells, once eliminated, cannot regenerate, because somatic cells, under received wisdom from Weismann, cannot make new germ line cells. Policymakers assume that the germ line cells are unique and have made decisions, for example, to allow genetic engineering of somatic cells under controlled circumstances but to prohibit engineering of germ line cells.

The work of MBL scientist Duygu Özpolat suggests otherwise. Her lab looks precisely at reproductive cell regeneration, a phenomenon perceived as impossible without experimental intervention, such as the work discussed above (https://www.mbl.edu/bell/current-faculty/duygu-ozpolat/, accessed August 22, 2020). She has found that in marine annelids, such as *Pristina leidyi*, reproductive cells are eliminated under stress conditions, like starvation, and then regenerate naturally.[14] In fact, germ line cell regeneration has been found to occur broadly in metazoans, including by converting somatic cells into germ line cells.[15] This evidence about the widespread nature of germ line cell regeneration has led Özpolat and her collaborators, including Kate MacCord, to ask: what are the regulatory processes guiding germ cell identity and regeneration? Understanding how cells become germ line cells and what their regenerative capabilities are will show a great deal about how the system undergoes regeneration and remains functional even while individual cellular parts change.

These are just a few among many examples of recent research on regeneration in organisms. Additional work looks at those scales of living systems outside the organism itself. In particular, ecosystems and microbial communities include many individuals, from diverse species, interacting in complex ways to make up a whole functioning system. We look at these scales and then return to ask to what extent understanding of regeneration across these different living systems reflects an underlying logic of regeneration. And how can we draw on that logic to inform our understanding of systems damage or injury at additional scales of living systems and ultimately the planet as a whole?

Ecosystems are another level of living system that undergoes cycles of growth, damage, and repair. Historians of ecology have chosen different points as the origin of ecological science, but here we start with British botanist Arthur Tansley, who is widely credited with having coined the term "ecosystem" in 1935. His summary article, "The Use and Abuse of Vegetative Concepts and Terms," appeared in the journal *Ecology* and examined the terms succession, development, climaxes, "complex organism," and then introduced "ecosystem." Henry Chandler Cowles and Frederic Clements had laid out ideas of succession and development in plant communities beginning in 1916. Clements especially developed an idea of the vegetative community as developing, like an organism, and previous studies of organic development and life cycles provided a guide for understanding progressive stages. Forests and other units reach a climax when they are fully developed through succession of different stages, and they become more complex throughout the process.

Tansley found this borrowing of concepts from individual development limiting. Rather than seeing units of ecological study as organisms, he saw them as "quasi-organisms." Yes, these ecological units have many features that Clements could construe as like those of organisms. But, Tansley argued, ecological biotic systems are not really the same as organisms. "There is no need to weary the reader with a list of the points in which the biotic community does *not* resemble the single animal or plant," he wrote. "They are so obvious and so numerous that the dissent expressed and even the ridicule poured on the proposition that vegetation *is* an organism are easily understood." Tansley noted that, even so, Clements insisted that a plant community is really an organism, not exactly an *individual organism,* but rather a "complex organism" made up of many different organisms.[16] Tansley rejected this attempt and insisted that plant communities are organisms only in a "quasi" sense.

Instead, Tansley suggested the concept of ecosystem. He captured what we still mean today by a complex system: "a whole *system* (in the sense of physics), including not only the organism-complex, but also the whole complex of physical factors forming what we call the environment of the biome — the habitat factors in the widest sense." He also recognized that humans are an active part of ecosystems and must be factored in.[17] This is a very clear statement that the ecosystem consists of all the living organisms and the environment, interacting and connected together. Tansley recognized that the whole is not neatly bounded, but neither is an individual organism. These ecosystems also go through developmental stages and therefore behave as quasi organisms but not exactly as individual organisms. In laying out these concepts so clearly, Tansley did not persuade everyone. But he did establish a strong foundation on which ecosystems ecologists could build. He insisted on the importance of the system, but he did not go on to tell us what methods would allow effective study of that system.

Three decades later, in 1966, systems ecologist George Van Dyne offered a report through the Oak Ridge National Laboratory (ORNL) looking at the intersection of "Ecosystems, Systems Ecology, and Systems Ecologists." There he suggested what is involved in studying systems and developing

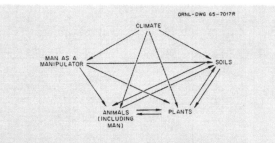

Fig. 1. An ecosystem is an integrated complex of living and nonliving components. Each component is influenced by the others, with the possible exception of macroclimate. And now man is on the verge of exerting meaningful influence over macroclimate.

FIGURE 4.2 | Figure 1 from George M. Van Dyne, "Ecosystems, Systems Ecology, and Systems Ecologists," 1966 Report from Oak Ridge National Laboratory.

models to allow explanation and prediction (figs. 4.2 and 4.3). He explained that the Radiation Ecology Section of ORNL's Health Physics Division had begun a new program on "Systems Ecology." This was a new field, greeted with skepticism by those who had succeeded under old regimes in ecology and with some support by those who were gaining from the new approach. So, what was new?

Systems ecology looked at renewable resource management, drawing on the contributions of ecologists and computation analysis. Just two years before, ecologist Eugene P. Odum had called for "The New Ecology" that would build on historical studies but now focused on ecosystems as the basic unit for analysis, just as cells or molecules served as basic units within other living systems. The ecosystem depends on communication among the parts and factors, with regulation of the whole system. And that system includes a diversity of different species, so it is different in that respect from the way in which

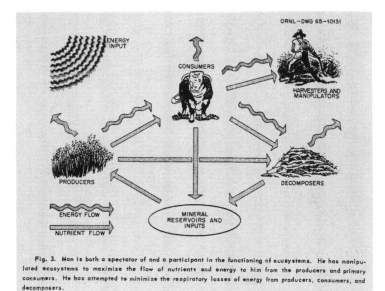

Fig. 3. Man is both a spectator of and a participant in the functioning of ecosystems. He has manipulated ecosystems to maximize the flow of nutrients and energy to him from the producers and primary consumers. He has attempted to minimize the respiratory losses of energy from producers, consumers, and decomposers.

FIGURE 4.3 | Figure 3 from George M. Van Dyne, "Ecosystems, Systems Ecology, and Systems Ecologists," 1966 Report from Oak Ridge National Laboratory.

a cell or an individual organism is a system. Odum urged ecologists to work together to study these systems, using mathematics and working toward developing models for explanation and prediction. Such coordination would be necessary to make ecological knowledge useful for policymakers and resource managers, as well as for the scientists themselves, he urged.[18]

Van Dyne had emphasized this need for mathematical modeling. Computers would prove valuable in drawing together vast amounts of data of diverse sorts, he urged. Scientists from different fields must work together, sharing access to computing power, training graduate students, and generating knowledge to be used for decision-making. This theme, that ecological knowledge was essential for effective resource management, gained further emphasis through the 1960s and 1970s, and again today in the context of the kinds of climate change that threaten failure of our planetary system. As ecosystems ecologists embraced computational and interdisciplinary approaches, they also recognized that ecosystems grow through stages of development, mature, and also suffer injury or damage. In some cases, the systems collapse with widespread failure. Examples became apparent: forest loss from insect infestations or extensive wildfires, soil degradation from excessive use of pesticides and herbicides that killed microbial communities necessary for agricultural health, water system contamination and impacts of the biotic communities, and so many more. Systems suffer, experience damage, and can even fail. Therefore, knowledge about how the parts of the system interact and regulate health can support wise management and controlling life, as Loeb put it, to restore ecosystem health. That was the goal.

Restoration ecology was oriented toward restoring health following disturbance, with a focus on recovering ecosystem function rather than restoring all of the parts of the system. Restoration should bring back a system we consider healthy because it "works" as a functioning whole system, but it might not be identical to the one that underwent damage, and in addition ecosystems evolve in ways that should make them different over time. As ecologist James P. Collins has explained, evolution became a central part of ecology through study of population changes, both populations of particular species and relative populations of different species. This emphasis brought genetics to ecology as well.[19] Thus, studying ecosystems drew on computational tools along with evolution, genetics, and modeling.

In more recent decades restoration ecology has been joined by discussions of resilience and eventually regeneration in ecosystems. Restoration, resilience, and replacement all have slightly different meanings, but all are useful for understanding complex systems such as ecosystems.

Traditional ecosystems ecology focused on macrobial, or macroscopic, organisms like salmon, maple trees, ants, or gophers. But microbial organisms, like bacteria, are also a part of life, and researchers have begun to realize that they are even essential for life. In agriculture, crops fail to thrive or die without microbes in the soil. Forests decline and die without microbes. Many animals, such as mice or bumble bees, show reduced body weight, increased susceptibility to pathogens, and increased mortality when they are raised in environments free of microbes. This realization that microbes are essential, rather than nasty invaders, represents a radical change in our attitude toward them: the twentieth century brought a gradual realization that microbes are not all bad; they are not all the pathogens that cause diseases and that Louis Pasteur, Robert Koch, and the germ theory had sought to eradicate. We've found that, in fact, "the microbial world constitutes the life support system of the biosphere," as it was put in a recent *Nature* consensus statement.[20]

This changing perspective entices us to find ways to grapple with microbial regeneration. How do we make sure that the microbes required for all living systems continue to regenerate successfully? At first, the question was: which particular microbes support the health of a system? But gradually, research has suggested that we need to focus on understanding communities of microbes. This has, in turn, suggested that the key to understanding microbial regeneration will lie in understanding the ecology and evolution responsible for how communities of microbes develop through time and respond to damage.

Although we can make use of traditional ecological and evolutionary theories, recent research has also shown that understanding microbial communities will require the development of new ways of thinking about systems and systems change. For example, W. Ford Doolittle and S. Andrew Inkpen argue—in their "It's the Song, Not the Singer" theory—that we should pay increased attention to the way that community functions themselves evolve through time and that these "songs," rather than the "singers" (microbes themselves) that perform them, are legitimate units of evolution by natural selection.[21] What are the microbes doing, and more precisely, what is the community of interacting microbes doing? Such thinking calls for a systems approach like that urged by the ecologists, and it reveals a set of rules guiding the regenerative processes.

A living system needs a microbial community, whether it is an ecosystem such as a forest or agricultural field, or an animal or plant that relies on microbes for digestion and other functions. The concept of a microbiome dates back long before the 2001 articulation by Joshua Lederberg that is often cited as the beginning. Yet, undoubtedly, the idea of an integrated microbiome made up of multiple different species of microbes serving together in conjunction with organisms they inhabit for digestion and other functions is relatively recent.[22]

Microbial communities are living systems that can grow, develop, and evolve. They can also experience damage, for example, when the gut microbiome is assaulted with a heavy dose of antibiotics. The antibiotics wipe out the biota in the gut. A form of replacement or regeneration occurs when "probiotics" allow a microbial community to recolonize. It is not likely to be exactly the same community and not likely to have all the identical species. But if it can "play the song," then it works. Regeneration restores the function and the system as a whole. This is also not simply a problem for individual animals or plants but also for all scales of life up to the whole biosphere. As infectious disease expert Martin Blaser has argued, the overuse of antibiotics in medicine and industrial agriculture has had the unintended consequence of altering the microbial communities throughout our global environment in ways that are not in harmony with the lasting health of many current biological systems worldwide.[23] Sustainability involves understanding the regeneration of healthful microbial communities.

The end goal of thinking of regeneration as a phenomenon within — and across — all living systems is to generate knowledge that can be transferred and translated among different scales of living systems in order to improve the health of our bodies, our ecosystems, and our planet. This systems-based approach, steeped in understanding the logic of regeneration within and across systems, holds enormous promise. What do we need in order to reach this promise?

We need to parse out the logic of regeneration within living systems by understanding the sets of rules guiding regenerative processes, to develop models that make knowledge translatable across the scales of living systems, and to develop tools that help us understand the interconnections among and across scales. To reach this point, we need greater clarity about what kinds of variables should be included, how they will be measured, how the data will be aggregated, and how summaries of descriptions can lead to explanations and predictions. We will need to look across scientific disciplines to achieve this and rely on history and philosophy to guide us. So far, we have a few models like Eric Davidson's that might be translatable to apply to all scales of living systems. Yet the detail of his models comes from decades of close study of sea urchins. We have some ecosystems models, but also considerable negotiation about what should count as relevant variables within the models. Some ecosystems ecologists do not include humans, for example, or only as biological units rather than also as social actors. We have a long way to go to achieve anything close to the control of life that Loeb envisioned.

Controlling life, as Loeb envisioned, and reengineering our bodies and our planet into health also depends on understanding the limits of living systems. Some systems become so damaged that they fail. Many are arguing—often quite vehemently and with increasing urgency—that our planetary system is in danger of failing in this era of the Anthropocene, in which humans dominate and often not wisely. We get climate change as a result, without sufficient understanding of the complex coupled natural-social system so that we know how to engineer solutions. Perhaps humans have achieved some examples of the kind of control and ability to engineer life that Loeb envisioned, except we do not really know what we are doing. Perhaps in some cases we are actually damaging the system further. Too many antibiotics can make it difficult for individual humans to recover. Too many pesticides or herbicides can destroy soil fertility. Too much wildfire caused by human intervention to suppress natural fire over time can scorch and sterilize the forest beyond repair. To truly understand how to harness regeneration in order to heal our bodies and our planet, we need to know how far a system can bend before it breaks, and then what to do if it breaks. A broken living system does not necessarily mean death; if a human does not regenerate a spinal neuron, they do not die. If an ecosystem or a microbial community cannot replace lost parts and regain function, they do not necessarily die. They can still exist, but they do so in an altered state. We need much better understanding of the processes involved in regenerating and restoring the systems involved.

All of this is going to take a great deal of thoughtful effort from people working across disciplines, from scientists to philosophers, from computational modelers to historians. We cannot rely solely on one discipline or one initiative to solve this problem, we need a concerted effort to move us closer to a broad and actionable understanding of regeneration. We need informed readers to help think about how we can understand regeneration and how we can use that knowledge.

# Acknowledgments

We would not have started this project if Susan Fitzpatrick hadn't challenged us to think differently, particularly in the context of the Marine Biological Laboratory. A relatively small institution for research and education, the MBL features three central, traditional areas of focus: neurobiology and cell biology, microbial evolution, and ecosystems ecology. People tend to reside in one or another of these groups, so Susan asked whether we could find anything in common. Could we identify ways to work across the different subdisciplines and specializations within biology?

That led to grants from the James S. McDonnell Foundation, for which Susan has served as president since 2016. Then-Director of the MBL Hunt Willard and Jane Maienschein outlined a plan and brought Kate MacCord in as coordinator. MacCord became codirector of subsequent grants as we focused in on regeneration. Current MBL Director Nipam Patel has been an enthusiastic supporter of the project, as has MBL Research Director David Mark Welch. A number of MBL researchers have participated in the workshops and are members of our working groups, and Director of the Eugene Bell Center for Regenerative Biology and Tissue Engineering Jennifer Morgan is a leader in our neurobiology working group. Numerous discussions at the MBL, especially back in the days before COVID when we were able to gather informally in person, introduced new suggestions. As the director of MBL development noted, looking at regeneration is exactly the kind of cross-cutting project that shows what an institution like the MBL makes possible because of its size, the ways people can interact, and the mix of research and education missions.

One of those research-education initiatives is the annual History of Biology seminar, funded by Arizona State University (ASU) through the Center for Biology and Society in collaboration with the MBL's informal education program. Co-organized by Jane Maienschein, John Beatty (at University of British Columbia), James Collins (at ASU), and Karl Matlin (at the University of Chicago and the MBL), the seminar is coordinated by Center Assistant Director Jessica Ranney. In 2018, we brought together an international group to work for a week with MBL researchers exploring questions about regeneration.

Our work builds on our experience with the MBL History Project, which has led to a digital archive of materials related to the MBL and history of biology: http://history.archives.mbl .edu. For that project, MBL librarians Jen Walton and Matt Person have been especially important, with technical support from John Furfey and encouragement from Nancy Stafford.

The core team of working group leaders includes Kathryn Maxson Jones and Jennifer Morgan for neuroscience,

Lucie Laplane and Michel Vervoort for stem cells, Kate Mac-Cord and B. Duygu Özpolat for germ lines, S. Andrew Inkpen and W. Ford Doolittle for microbial evolution, and Frederick Davis and James Collins for ecosystems ecology. These group leaders meet regularly, read each other's work, exchange ideas, argue about how best to capture shared interpretations, and also acknowledge how differently philosophers, historians, and biologists all think about things. Kathryn Maxson Jones coordinates us all and keeps us on task. We have a lot of fun and are living proof that working across disciplinary boundaries may be more work, but it is definitely worth it for thinking more deeply and clearly. Where one or another of us initially thought "that doesn't make any sense, we don't think that way," now we say "oh, is this what you mean by that? Might that be the same as what I mean by this-other-thing?" And so on.

After all the discussions comes the writing. Here we thank every member of our working team for reading and rereading and offering valuable suggestions. In addition, we thank Richard Creath and Challie Facemire for their continued support. Friends Eric Ullman, Manfred Laubichler, and Hanna Worliczek read parts or all of the manuscript and offered valuable suggestions. And our editor at the University of Chicago Press, Joseph Calamia, has consistently offered exceptionally helpful suggestions, including about what did not make sense or was not explained sufficiently for a general reader, and always in the most generous and positive way. This book is really a team project, as are the other books that will appear in this series of small regeneration explorations.

# Notes

## CHAPTER 1

1   Paul A. Oliphint et al., "Regenerated Synapses in Lamprey Spinal Cord Are Sparse and Small even after Functional Recovery from Injury," *Journal of Comparative Neurology* 518 (2010): 2854–72.

2   Ruth Lehmann, ed., *The Immortal Germline* (New York: Elsevier, 2019); Chris Smelick and Shawn Ahmed, "Achieving Immortality in the C. elegans Germline," *Ageing Research Reviews* 4 (2005): 67–82; Françoise Baylis, *Altered Inheritance: CRISPR and the Ethics of Human Genome Editing* (Cambridge, MA: Harvard University Press, 2019).

3   Lucie Laplane, *Cancer Stem Cells: Philosophy and Therapies* (Cambridge, MA: Harvard University Press, 2016).

## CHAPTER 2

1   Thomas Hunt Morgan, *Regeneration* (New York: Macmillan, 1901).

2   Andrea Falcon, "Aristotle on Causality," *Stanford Encyclopedia of Philosophy* (2019, first published 2006). https://plato.stanford.edu/entries/aristotle-causality/.

3   James Lennox, "Aristotle's Biology," *Stanford Encyclopedia of Philosophy* (2017, first published 2006). https://plato.stanford.edu/entries/aristotle-biology/.

4   Aristotle, *History of Animals*, trans. Richard Cresswell, *Aristotle's History of Animals in Ten Books* (London: George Bell and Sons, 1902), part 12, 44;

Jane Maienschein, *Embryos Under the Microscope: The Diverging Meanings of Life* (Cambridge, MA: Harvard University Press, 2014), 31–35.

5    Richard J. Goss, "The Natural History (and Mystery) of Regeneration," *A History of Regeneration Research: Milestones in the Evolution of a Science*, ed. Charles E. Dinsmore (Cambridge: Cambridge University Press, 1991), 9–12.

6    Conrad Gesner, *Historiae Anima* (Zurich: Apvd Christ. Froschovervm, 1551).

7    Paul Lawrence Farber, *Finding Order in Nature: The Naturalist Tradition from Linnaeus to E. O. Wilson* (Baltimore: Johns Hopkins University Press, 2000).

8    D. R. Newth, "New (and Better?) Parts for Old," in *New Biology*, edited by M. L. Johnson, M. Abercrombie, and G. E. Fogg, 47–48 (London: Penguin Books, 1958).

9    Shirley A. Roe, *Matter, Life, and Generation: 18th-Century Embryology and the Haller-Wolff Debate* (Cambridge: Cambridge University Press, 1981).

10   Jane Maienschein, *Whose View of Life? Embryos, Cloning, and Stem Cells* (Cambridge, MA: Harvard University Press, 2005).

11   Mary Terrall, *Catching Nature in the Act: Réaumur and the Practice of Natural History in the Eighteenth Century* (Chicago: University of Chicago Press, 2014).

12   Terrall, 47.

13   René-Antoine Ferchault de Réaumur, "Sur les Diverses Reproductions qui se font dans les Ecrevisse, les Omars, les Crabes, etc. Et entr'autres sur celles de leurs Jambes et de leurs Écailles," *Memoires de l'Academie Royale des Sciences* (1712): 223–45.

14   Dorothy M. Skinner and John S. Cook, "New Limbs for Old: Some Highlights in the History of Regeneration in Crustacea," in *A History of Regeneration Research: Milestones in the Evolution of a Science*, ed. Charles E. Dinsmore (Cambridge: Cambridge University Press, 1991), 25–45.

15   Howard M. Lenhoff and Sylvia G. Lenhoff, "Trembley's Polyps," *Scientific American* 258 (1988): 108.

16   Howard M. Lenhoff and Sylvia G. Lenhoff, "Abraham Trembley and the Origins of Research on Regeneration in Animals," in *A History of Regeneration Research: Milestones in the Evolution of a Science*, ed. Charles E. Dinsmore (Cambridge: Cambridge University Press, 1991), 47–66.

17   Roe, 10, quoting Trembley 1744.

18   Abraham Trembley, *Mémoirs pour servir à l'histoire d'un genre de polypes d'eau douce, à bras en forme de cornes* (Leiden: Verbeck, 1744), trans. Howard M. Lenhoff and Sylvia G. Lenhoff, *Hydra and the Birth of Experimental*

*Biology, 1744: Abraham Trembley's Memoirs Concerning the Natural History of a Type of Freshwater Polyp with Arms Shaped Like Horns* (Pacific Grove, CA: Boxwood Press, 1986), 9.

19   Howard M. Lenhoff and Sylvia G. Lenhoff, "Abraham Trembley and the Origins of Research on Regeneration in Animals," in *A History of Regeneration Research: Milestones in the Evolution of a Science*, ed. Charles E. Dinsmore (Cambridge: Cambridge University Press, 1991), 62; Howard M. Lenhoff and Sylvia G. Lenhoff, "Challenge to the Specialist: Abraham Trembley's Approach to Research on the Organism—1744 and Today," *American Zoologist* 29 (1989): 1105–17. https://www.jstor.org/stable/3883509.

20   Charles Bonnet, *Oeuvres d'histoire naturelle et de philosophie*, 18 volumes (Neuchâtel: S Fauche, 1779–83).

21   Roe, 22–23.

22   Lazzaro Spallanzani, https://archive.org/stream/b30356167?ref=ol #mode/2up "Prodromo di un opera da imprimersi sopra la rirproduzioni animali." Modena: Giovanni Montanari (1768); trans. Matthew Maty, "An Essay on Animal Reproductions" (London: T. Becket and P. A. de Hondt, 1769).

23   Charles E. Dinsmore, "Lazzaro Spallanzani: Regeneration in Context," in *A History of Regeneration Research: Milestones in the Evolution of a Science*, ed. Charles E. Dinsmore (Cambridge: Cambridge University Press, 1991), 83.

24   Joseph A. Caron, "'Biology' in the Life Sciences: A Historiographical Contribution," *History of Science* 26 (1988): 223–68.

25   Frederick B. Churchill, "Wilhelm Roux and a Program for Embryology" (PhD diss. Harvard University, 1967); Jane Maienschein, "The Origins of *Entwicklungsmechanik*," in Scott Gilbert, ed., *A Conceptual History of Modern Embryology* (Cambridge: Cambridge University Press, 1991a), 43–61.

26   William Morton Wheeler, "Translation of Wilhelm Roux's 'The Problems, Methods and Scope of Developmental Mechanics,'" *Biological Lectures of the Marine Biological Laboratory* (Woods Hole, 1895), 149–90.

27   Maienschein, 1991a.

28   Frederick B. Churchill, *Weismann: Development, Heredity, and Evolution* (Cambridge, MA: Harvard University Press, 2015).

29   Wilhelm Roux, "Beiträge zur Entwickelungsmechanik des Embryo. Über die künstliche Hervorbringung halber Embryonen durch Zerstörung einer der beiden ersten Furchungskugeln, sowie über die Nachentwickelung (Postgeneration) der fehlenden Körperhälfte," *Virchows Archiv für Pathologische Anatomie und Physiologie und für Klinische Medizin* 114 (1888): 113–

53. Translated as "Contributions to the Development of the Embryo. On the Artificial Production of One of the First Two Blastomeres, and the Later Development (Postgeneration) of the Missing Half of the Body," in *Foundations of Experimental Embryology*, eds. Benjamin H. Willier and Jane M. Oppenheimer (New York: Hafner Press, 1964), 2–37.

30    Hans Driesch, "Entwicklungsmechanische Studien: I. Der Werthe der beiden ersten Furchungszellen in der Echinogdermenentwicklung. Experimentelle Erzeugung von Theil- und Doppelbildungen. II. Über die Beziehungen des Lichtez zur ersten Etappe der thierischen Form-bildung," *Zeitschrift für wissenschaftliche Zoologie* 53 (1891): 160–84. Translated as "The Potency of the First Two Cleavage Cells in Echinoderm Development. Experimental Production of Partial and Double Formations," in *Foundations of Experimental Embryology*, eds. Benjamin H. Willier and Jane M. Oppenheimer (New York: Hafner Press, 1964), 38–50.

31    Driesch.

32    Frederick B. Churchill, "Regeneration, 1885–1901," in *A History of Regeneration Research: Milestones in the Evolution of a Science*, ed. Charles E. Dinsmore (Cambridge: Cambridge University Press, 1991), 113.

33    Mary E. Sunderland, "Regeneration: Thomas Hunt Morgan's Window into Development," *Journal of the History of Biology* 43 (2010): 325–61.

34    Jane Maienschein, "T. H. Morgan's Regeneration, Epigenesis, and (W)holism," in *A History of Regeneration Research: Milestones in the Evolution of a Science*, ed. Charles E. Dinsmore (Cambridge: Cambridge University Press, 1991b), 133–49.

35    Thomas Hunt Morgan, *Regeneration* (New York: Macmillan, 1901), 278–79.

## CHAPTER 3

1    Jane Maienschein, *Transforming Traditions in American Biology, 1880–1915* (Baltimore: Johns Hopkins University Press, 1991c); Keith R. Benson, Jane Maienschein, and Ronald Rainger, eds. *The Expansion of American Biology* (New Brunswick, NJ: Rutgers University Press, 1991); Ronald Rainger, Keith R. Benson, and Jane Maienschein, *The American Development of Biology* (Philadelphia: University of Pennsylvania Press, 1988); republished in paperback (New Brunswick, NJ: Rutgers University Press, 1991).

2    Thomas Hunt Morgan, *The Development of the Frog's Egg: An Introduction to Experimental Embryology* (New York: Macmillan, 1897).

3    Thomas Hunt Morgan, *Embryology and Genetics* (New York: Columbia University Press, 1934); Jane Maienschein, "Garland Allen, Thomas Hunt

Morgan, and Development," *Journal of the History of Biology* 49 (2015): 587–601.

4    Alfred H. Sturtevant, "Thomas Hunt Morgan," *Biographical Memoirs of the National Academy of Sciences* 33 (1959): 283–325.

5    Thomas Hunt Morgan, *Regeneration* (New York: Macmillan, 1901), vii–viii.

6    Marga Vicedo, "T. H. Morgan: Neither an Epistemological Empiricist nor a 'Methodological Empiricist,'" *Biology and Philosophy* 5 (1990): 293–311; Nils Roll-Hansen, "Drosophila Genetics: A Reductionist Research Program," *Journal of the History of Biology* 11 (1978): 159–210.

7    Morgan, 1901, 255–56.

8    Morgan, 274.

9    Winthrop John Van Leuven Osterhout, "Jacques Loeb, 1859–1924," *Biographical Memoirs of the National Academy of Sciences* 13 (1930): 318–401. Reprinted from the Jacques Loeb Memorial Volume, *The Journal of General Physiology* VIII (no. 1, September 15, 1928): ix–xcii; Philip Pauly, *Controlling Life: Jacques Loeb and the Engineering Ideal in Biology* (New York: Oxford University Press, 1987).

10    Pauly, 55.

11    On biology at the University of Chicago, see Gregg Mitman and Adele E. Clarke, *Crossing the Borderlands: Biology at Chicago,* special issue of *Perspectives on Science* (Chicago: University of Chicago Press, 1993).

12    Pauly, 95.

13    Sturtevant, 288.

14    Pauly, 101–5.

15    Jacques Loeb, "On the Chemical Character of the Process of Fertilization and Its Bearing on the Theory of Life Phenomena," *Science* 26 (1907):425.

16    Loeb, "On the Chemical Character," 437.

17    Pauly, 148.

18    Jacques Loeb, *Mechanistic Conception of Life* (Chicago: University of Chicago Press, 1912), 3.

19    MBL Report for 1924, 25.

20    Jacques Loeb, *The Organism as a Whole* (New York: Putnam and Sons, 1916), chap. 7.

21    Osterhout, ix–xcii.

22    Loeb, *Regeneration.*

23    Loeb, *Regeneration,* 6.

24    Loeb, *Regeneration,* 8.

25    John W. Boyer, *The University of Chicago: A History* (Chicago: University of Chicago Press, 2015).

26    Libbie Hyman, "Charles Manning Child 1869–1954" in *Biographical Mem-*

*oirs of the National Academy of Sciences* (Washington: National Academy of Sciences, 1957), 73–103.

27    Maienschein, 1991c, 133.

28    Charles Manning Child, *Individuality in Organisms* (Chicago: University of Chicago Press, 1915a), x.

29    Child, *Individuality*, x.

30    Child, 125.

31    Child, 87.

32    Lewis Wolpert, "Morgan's Ambivalence: Gradients and Regeneration," in *A History of Regeneration Research: Milestones in the Evolution of a Science*, ed. Charles E. Dinsmore (Cambridge: Cambridge University Press, 1991), 215.

33    Richard A. Liversage, "Origin of the Blastema Cells in Epimorphic Regeneration of Urodele Appendages: A History of Ideas," in *A History of Regeneration Research: Milestones in the Evolution of a Science*, ed. Charles E. Dinsmore (Cambridge: Cambridge University Press, 1991), 179–99.

34    Jane Maienschein, "Ross Granville Harrison (1870–1959) and Perspectives on Regeneration," *Journal of Experimental Zoology B* 314 (2010): 607–15.

35    Magdalena Zernicka-Goetz and Roger Highfield, *The Dance of Life: The New Science of How a Single Cell Becomes a Human Being* (New York: Basic Books, 2020), 182.

36    Andrew R. Gehrke et al., "Acoel Genome Reveals the Regulatory Landscape of Whole-body Regeneration," *Science* 363 (2019): eaau6173. DOI: 10.1126/science.aau6173.

## CHAPTER 4

1    Anthony Trewavas, "A Brief History of Systems Biology: 'Every Object that Biology Studies Is a System of Systems,' Francois Jacob (1974)." *Plant Cell* 18 (2006): 2420–30.

2    Christopher Wanjek, "Systems Biology as Defined by NIH: An Intellectual Resource for Integrative Biology" (2016): https://irp.nih.gov/catalyst/v19i6/systems-biology-as-defined-by-nih (accessed 9 August 2020).

3    Arthur B. Pardee, François Jacob, and Jacques Monod, "The Role of the Inducible Alleles and the Constitutive Alleles in the Synthesis of Beta-galactosidase in Zygotes of *Escherichia coli*," *Comptes rendus hebdomadaires des seances de l'Academie des sciences* 246 (1958): 3125–28; Roy J. Britten and Eric H. Davidson, "Gene Regulation for Higher Cells: A Theory," *Science* 165 (1969): 349–57.

4     Sarah A. Elliott and Alejandro Sánchez Alvarado, "Planarians and the History of Animal Regeneration: Paradigm Shifts and Key Concepts in Biology," *Methods in Molecular Biology* 1774 (2018): 207–39.

5     Alejandro Sánchez Alvarado, "To Solve Old Problems, Study New Research Organisms," *Developmental Biology* 433 (2018): 111–14.

6     See Maienschein, *Whose View of Life?* and Jane Maienschein, *Embryos Under the Microscope: The Diverging Meanings of Life* (Cambridge, MA: Harvard University Press, 2014) on the history of stem cells, and Laplane, *Cancer Stem Cells*, on the philosophy of stem cells.

7     Nicholas Wade, "Scientists Cultivate Cells at Root of Human Life," *The New York Times*, November 6, 1998: https://www.nytimes.com/1998/11/06/us/scientists-cultivate-cells-at-root-of-human-life.html.

8     Marina Cavazzana-Calvo et al., "Gene Therapy of Human Severe Combined Immunodeficiency (SCID)-X1 disease," *Science* 288 (2000): 669–72.

9     Donald B. Kohn, "Gene Therapy for XSCID: The First Success of Gene Therapy," *Pediatric Research* 48, no. 5 (2000); Donald B. Kohn, Michel Sadelain, and Joseph C. Glorioso, "Occurrence of Leukaemia Following Gene Therapy of X-linked SCID," *Nature Reviews Cancer* 3 (2003): 477–88.

10     Santiago Ramón y Cajal, *Degeneration and Regeneration of the Nervous System* (London: Oxford University Press, 1928).

11     Hanna Valli et al., "Germline Stem Cells: Toward the Regeneration of Spermatogenesis," *Fertility and Sterility* 101, no. 1 (2014): 3–13.

12     Katsuhiko Hayashi et al., "Reconstitution of the Mouse Germ Cell Specification Pathway in Culture by Pluripotent Stem Cells," *Cell* 146 (2011): 519–32; Katsuhiko Hayishi and Mitinori Saitou, "Generation of Eggs from Mouse Embryonic Stem Cells and Induced Pluripotent Stem Cells," *Nature Protocols* 8, no. 8 (2013): 1513; Taichi Akahori, Dori C. Woods, and Jonathan L. Tilly, "Female Fertility Preservation Through Stem Cell–based Ovarian Tissue Reconstitution in vitro and Ovarian Regeneration in vivo," *Clinical Medicine Insights: Reproductive Health* 13 (2019): 1179558119848007.

13     Kate MacCord and B. Duygu Özpolat, "Is the Germline Immortal and Continuous? A Discussion in Light of iPSCs and Germline Regeneration," *Zenodo* (September 2019): doi:10.5281/zenodo.3385322.

14     B. Duygu Özpolat et al., "Plasticity and Regeneration of Gonads in the Annelid *Pristina leidyi*," *EvoDevo* 7 (2016): 1–15.

15     Keita Yoshida, et al., "Germ Cell Regeneration-mediated, Enhanced Mutagenesis in the Ascidian Ciona intestinalis Reveals Flexible Germ Cell Formation from Different Somatic Cells," *Developmental Biology* 423, no. 2 (2017): 111–25; Yuying Wang et al., "Nanos Function Is Essential for Development and Regeneration of Planarian Germ Cells," *Proceedings of the Na-*

*tional Academy of Sciences* 104 no. 14 (2007): 5901–6; Leah C. Dannenberg and Elaine C. Seaver, "Regeneration of the Germline in the Annelid Capitella teleta," *Developmental Biology* 440, no. 2 (2018): 74–87; Angela N. Kaczmarczyk, "Germline Maintenance and Regeneration in the Amphipod Crustacean, Parhyale hawaiensis," PhD diss., University of California Berkeley (2014).

16    Arthur Tansley, "The Use and Abuse of Vegetation Concepts and Terms," *Ecology* 156 (1935): 290–91.

17    Tansley, 299.

18    Eugene P. Odum, "The New Ecology," *Bioscience* 14 (1964): 14–16.

19    James P. Collins, "'Evolutionary Ecology' and the Use of Natural Selection in Ecological Theory," *Journal of the History of Biology* 19 (1986): 257–88.

20    Richard Cavicchioli et al., "Scientists' Warning to Humanity: Microorganisms and Climate Change," *Nature Reviews Microbiology* 17 (2019): 569–86.

21    W. Ford Doolittle and S. Andrew Inkpen, "Processes and Patterns of Interaction as Units of Selection: An introduction to ITSNTS Thinking," *Proceedings of the National Academy of Sciences* 115 (2018): 4006–14; first published March 26, 2018: https://doi.org/10.1073/pnas.1722232115.

22    For history from the NIH perspective, see: https://commonfund.nih.gov/hmp; also Kenneth D. Aiello, "Systematic Analysis of the Factors Contributing to the Variation and Change of the Microbiome," PhD diss., Arizona State University (2018), and Kenneth D. Aiello and Michael Simeone, "Triangulation of History Using Textual Data," *ISIS* 110 (2019): 522–37. https://doi.org/10.1086/705541.

23    Martin J. Blaser, *Missing Microbes: How the Overuse of Antibiotics Is Fueling Our Modern Plagues* (New York: Henry Holt and Company, 2014).

# References

Aiello, Kenneth D. 2018. "Systematic Analysis of the Factors Contributing to the Variation and Change of the Microbiome." PhD diss., Arizona State University.

Aiello, Kenneth D., and Michael Simeone. 2019. "Triangulation of History Using Textual Data." *ISIS* 110: 522–37. https://doi.org /10.1086/705541.

Akahori, Taichi, Dori C. Woods, and Jonathan L. Tilly. 2019. "Female Fertility Preservation through Stem Cell–based Ovarian Tissue Reconstitution in vitro and Ovarian Regeneration in vivo." *Clinical Medicine Insights: Reproductive Health* 13: 1179558119848007.

Aristotle. 1902. *History of Animals.* Translated by Richard Cresswell, *Aristotle's History of Animals in Ten Books.* London: George Bell and Sons.

Baylis, Françoise. 2019. *Altered Inheritance: CRISPR and the Ethics of Human Genome Editing.* Harvard University Press.

Benson, Keith R., Jane Maienschein, and Ronald Rainger. 1991. *The Expansion of American Biology.* New Brunswick, NJ: Rutgers University Press.

Blaser, Martin J. 2014. *Missing Microbes: How the Overuse of Antibiotics Is Fueling Our Modern Plagues.* New York: Henry Holt and Company.

Bonnet, Charles. 1779–83. *Oeuvres d'histoire naturelle et de philosophie.* 18 vols. Neuchâtel: S Fauche.

Boyer, John W. 2015. *The University of Chicago: A History.* Chicago: University of Chicago Press.

Britten, Roy J., and Eric H. Davidson. 1969. "Gene Regulation for Higher Cells: A Theory." *Science* 165: 349–57.

Caron, Joseph A. 1988. "'Biology' in the Life Sciences: A Historiographical Contribution." *History of Science* 26: 223–68.

Cavazzana-Calvo, Marina, Salima Hacein-Bey, Geneviève de Saint Basile, Fabian Gross, Eric Yvon, Patrick Nusbaum, Françoise Selz, Christophe Hue, Stéphanie Certain, Jean Laurent Casanova, Philippe Bousso, Françoise Le Deist, and Alain Fischer. 2000. "Gene Therapy of Human Severe Combined Immunodeficiency (SCID)-X1 Disease." *Science* 288: 669–72.

Cavicchioli, Richard, William J. Ripple, Kenneth N. Timmis, Farooq Azam, Lars R. Bakken, Matthew Baylis, Michael J. Behrenfeld, Antje Boetius, Philip W. Boyd, Aimée T. Classen, Thomas W. Crowther, Roberto Danovaro, Christine M. Foreman, Jef Huisman, David A. Hutchins, Janet K. Jansson, David M. Karl, Britt Koskella, David B. Mark Welch, Jennifer B. H. Martiny, Mary Ann Moran, Victoria J. Orphan, David S. Reay, Justin V. Remais, Virginia I. Rich, Brajesh K. Singh, Lisa Y. Stein, Frank J. Stewart, Matthew B. Sullivan, Madeleine J. H. van Oppen, Scott C. Weaver, Eric A. Webb, and Nicole S. Webster. 2019. "Scientists' Warning to Humanity: Microorganisms and Climate Change," *Nature Reviews Microbiology* 17: 569–86.

Child, Charles Manning. 1915a. *Individuality in Organisms.* Chicago: University of Chicago Press.

Child, Charles Manning. 1915b. *Senescence and Rejuvenescence.* Chicago: University of Chicago Press.

Child, Charles Manning. 1941. *Patterns and Problems in Development.* Chicago: University of Chicago Press.

Churchill, Frederick B. 1967. "Wilhelm Roux and a Program for Embryology." Harvard University PhD diss.

Churchill, Frederick B. 1991. "Regeneration, 1885–1901." In *A History of Regeneration Research: Milestones in the Evolution of a Science,* edited by Charles E. Dinsmore, 113–31. Cambridge: Cambridge University Press.

Churchill, Frederick B. 2015. *Weismann: Development, Heredity, and Evolution.* Cambridge, MA: Harvard University Press.

Collins, James P. 1986. "'Evolutionary Ecology' and the Use of Natural Selection in Ecological Theory." *Journal of the History of Biology* 19:257–88.

Crowe, Nathan, Michael R. Dietrich, Beverly Alomepe, Amelia Antrim, Bay Lauris ByrneSim, and Yi He. 2015. "The Diversification of Developmental Biology," *Studies in History and Philosophy of Biological and Biomedical Sciences* 53: 1–15.

Dannenberg, Leah C., and Elaine C. Seaver. 2018. "Regeneration of the Germline in the Annelid Capitella teleta." *Developmental Biology* 440 (no. 2): 74–87.

Dinsmore, Charles E. 1991. "Lazzaro Spallanzani: Regeneration in Context." In *A History of Regeneration Research: Milestones in the Evolution of a Science,* edited by Charles E. Dinsmore, 67–89. Cambridge: Cambridge University Press.

Dinsmore, Charles E. 1991. *A History of Regeneration Research:*

*Milestones in the Evolution of a Science*. Cambridge: Cambridge University Press.

Doolittle, W. Ford, and S. Andrew Inkpen. 2018. "Processes and Patterns of Interaction as Units of Selection: An Introduction to ITSNTS Thinking." *Proceedings of the National Academy of Sciences* 115: 4006–14; first published March 26, 2018: https://doi.org/10.1073/pnas.1722232115.

Driesch, Hans. 1891. "Entwicklungsmechanische Studien: I. Der Werthe der beiden ersten Furchungszellen in der Echinogdermenentwicklung. Experimentelle Erzeugung von Theil- und Doppelbildungen. II. Über die Beziehungen des Lichtez zur ersten Etappe der thierischen Form-bildung." *Zeitschrift für wissenschaftliche Zoologie* 53: 160–84. Translated as "The Potency of the First Two Cleavage Cells in Echinoderm Development. Experimental Production of Partial and Double Formations." In *Foundations of Experimental Embryology*, edited by Benjamin H. Willier and Jane M. Oppenheimer, 38–50. New York: Hafner Press, 1964.

Elliott, Sarah A., and Alejandro Sánchez Alvarado. 2018. "Planarians and the History of Animal Regeneration: Paradigm Shifts and Key Concepts in Biology." *Methods in Molecular Biology* 1774: 207–39.

Falcon, Andrea. 2019. "Aristotle on Causality." *Stanford Encyclopedia of Philosophy*. First published 2006. https://plato.stanford.edu/entries/aristotle-causality/.

Farber, Paul Lawrence. 2000. *Finding Order in Nature: The Naturalist Tradition from Linnaeus to E. O. Wilson*. Baltimore: Johns Hopkins University Press.

Gehrke, Andrew R., Emily Neverett, Yi-Jyun Luo, Alexander

Brandt, Lorenzo Ricci, Ryan E. Hulett, Annika Gompers, J. Graham Ruby, Daniel S. Rokhsar, Peter W. Reddien, and Mansi Srivastava. 2019. "Acoel Genome Reveals the Regulatory Landscape of Whole-body Regeneration." *Science* 363: eaau6173. DOI: 10.1126/science.aau6173

Gesner, Conrad. 1551. *Historiae Anima*. Zurich: Apvd Christ. Froschovervm.

Goss, Richard J. 1991. "The Natural History (and Mystery) of Regeneration." In *A History of Regeneration Research: Milestones in the Evolution of a Science*, edited by Charles E. Dinsmore, 7–23. Cambridge: Cambridge University Press.

Hayashi, Katsuhiko, Hiroshi Ohta, Kazuki Kurimoto, Shinya Aramaki, and Mitinori Saitou. 2011. "Reconstitution of the Mouse Germ Cell Specification Pathway in Culture by Pluripotent Stem Cells." *Cell* 146: 519–32.

Hayashi, Katsuhiko, and Mitinori Saitou. 2013. "Generation of Eggs from Mouse Embryonic Stem Cells and Induced Pluripotent Stem Cells." *Nature Protocols* 8 (no. 8): 1513.

Hyman, Libbie. 1957. "Charles Manning Child 1869–1954." *Biographical Memoirs of the National Academy of Sciences*. Washington, DC: National Academy of Sciences, 73–103.

Kaczmarczyk, Angela N. 2014. "Germline Maintenance and Regeneration in the Amphipod Crustacean, Parhyale hawaiensis." PhD diss., UC Berkeley.

Kohn, Donald B. 2000. "Gene Therapy for XSCID: The First Success of Gene Therapy." *Pediatric Research* 48 (no. 5): 578.

Kohn, Donald B., Michel Sadelain, and Joseph C. Glorioso. 2003. "Occurrence of Leukaemia Following Gene Therapy of X-linked SCID." *Nature Reviews Cancer* 3: 477–88.

Laplane, Lucie. 2016. *Cancer Stem Cells: Philosophy and Therapies*. Cambridge, MA: Harvard University Press.

Lehmann, Ruth, ed. 2019. *The Immortal Germline*. New York: Elsevier.

Lenhoff, Howard M., and Sylvia G. Lenhoff. 1988. "Trembley's Polyps." *Scientific American* 258: 108–13.

Lenhoff, Howard M., and Sylvia G. Lenhoff. 1989. "Challenge to the Specialist: Abraham Trembley's Approach to Research on the Organism—1744 and Today." *American Zoologist* 29: 1105–17. https://www.jstor.org/stable/3883509.

Lenhoff, Howard M., and Sylvia G. Lenhoff. 1991."Abraham Trembley and the Origins of Research on Regeneration in Animals." In *A History of Regeneration Research: Milestones in the Evolution of a Science*, edited by Charles E. Dinsmore, 47–66. Cambridge: Cambridge University Press.

Lennox, James. 2017. "Aristotle's Biology." *Stanford Encyclopedia of Philosophy*. First published 2006. https://plato.stanford.edu/entries/aristotle-biology/.

Liversage, Richard A. 1991. "Origin of the Blastema Cells in Epimorphic Regeneration of Urodele Appendages: A History of Ideas." In *A History of Regeneration Research: Milestones in the Evolution of a Science*, edited by Charles E. Dinsmore, 179–99. Cambridge: Cambridge University Press.

Loeb, Jacques. 1907. "On the Chemical Character of the Process of Fertilization and Its Bearing on the Theory of Life Phenomena." *Science* 26: 425–37.

Loeb, Jacques. 1912. *Mechanistic Conception of Life*. Chicago: University of Chicago Press.

Loeb, Jacques. 1916. *The Organism as a Whole.* New York: Putnam and Sons.

Loeb, Jacques. 1924. *Regeneration.* New York: McGraw-Hill.

MacCord, Kate, and B. Duygu Özpolat. 2019. "Is the Germline Immortal and Continuous? A Discussion in Light of iPSCs and Germline Regeneration." *Zenodo* (September). doi:10.5281/zenodo.3385322.

Maienschein, Jane. 1991a. "The Origins of Entwicklungsmechanik," in *A Conceptual History of Modern Embryology*, edited by Scott Gilbert, 43–61. Cambridge: Cambridge University Press.

Maienschein, Jane. 1991b. "T. H. Morgan's Regeneration, Epigenesis, and (W)holism." In *A History of Regeneration Research: Milestones in the Evolution of a Science*, edited by Charles E. Dinsmore, 133–49. Cambridge: Cambridge University Press.

Maienschein, Jane. 1991c. *Transforming Traditions in American Biology, 1880–1915.* Baltimore: Johns Hopkins University Press.

Maienschein, Jane. 2005. *Whose View of Life? Embryos, Cloning, and Stem Cells.* Cambridge, MA: Harvard University Press.

Maienschein, Jane. 2010. "Ross Granville Harrison (1870–1959) and Perspectives on Regeneration," *Journal of Experimental Zoology B* 314: 607–15.

Maienschein, Jane. 2014. *Embryos Under the Microscope: The Diverging Meanings of Life.* Cambridge, MA: Harvard University Press.

Maienschein, Jane. 2015. "Garland Allen, Thomas Hunt Morgan, and Development." *Journal of the History of Biology* 49: 587–601.

Marine Biological Laboratory Annual Reports. 1925 for the year

1924. Available: https://hpsrepository.asu.edu/bitstream/handle/10776/1477/1924.pdf.

Mitman, Gregg, and Adele E. Clarke. 1993. "Crossing the Borderlands: Biology at Chicago," special issue of *Perspectives on Science*. Chicago: University of Chicago Press.

Morgan, Thomas Hunt. 1897. *The Development of the Frog's Egg: An Introduction to Experimental Embryology*. New York: MacMillan.

Morgan, Thomas Hunt. 1901. *Regeneration*. New York: Macmillan.

Morgan, Thomas Hunt. 1934. *Embryology and Genetics*. New York: Columbia University Press.

Newth, D. R. 1958. "New (and Better?) Parts for Old." In *New Biology*, edited by M. L. Johnson, M. Abercrombie, and G. E. Fogg. London: Penguin Books, 47–62.

Odum, Eugene P. 1964. "The New Ecology." *Bioscience* 14: 14–16.

Oliphint, Paul A., Naila Alieva, Andrea E. Foldes, Eric D. Tytell, Billy Y-B. Lau, Jenna S. Pariseau, Avis H. Cohen, and Jennifer R. Morgan. 2010. "Regenerated Synapses in Lamprey Spinal Cord Are Sparse and Small Even After Functional Recovery from Injury." *Journal of Comparative Neurology* 518: 2854–72.

Osterhout, Winthrop John Van Leuven. 1930. "Jacques Loeb, 1859–1924." *Biographical Memoirs of the National Academy of Sciences* 13: 318–401. Reprinted from the Jacques Loeb Memorial Volume. *The Journal of General Physiology* VIII (no. 1; September 15, 1928): ix–xcii.

Özpolat, B. Duygu, Emily S. Sloane, Eduardo E. Zattara, and Alexandra E. Bely. 2016. "Plasticity and Regeneration of Gonads in the Annelid Pristina leidyi." *EvoDevo* 7: 1–15.

Pardee, Arthur B., François Jacob, and Jacques Monod. 1958. "The Role of the Inducible Alleles and the Constitutive Alleles in the

Synthesis of Beta-galactosidase in Zygotes of *Escherichia coli.*"
*Comptes rendus hebdomadaires des seances de l'Academie des sci-*
*ences* 246: 3125–28.

Pauly, Philip. 1987. *Controlling Life: Jacques Loeb and the Engineering*
*Ideal in Biology.* New York: Oxford University Press.

Rainger, Ronald, Keith R. Benson, and Jane Maienschein. 1988.
*The American Development of Biology.* Philadelphia: University
of Pennsylvania Press; republished in paperback 1991. New
Brunswick, NJ: Rutgers University Press.

Ramón y Cajal, Santiago. 1928. *Degeneration and Regeneration of the*
*Nervous System.* London: Oxford University Press.

Réaumur, René-Antoine Ferchault de. 1712. "Sur les Diverses
Reproductions qui se font dans les Ecrevisse, les Omars, les
Crabes, etc. Et entr'autres sur celles de leurs Jambes et de
leurs Écailles." *Memoires de l'Academie Royale des Sciences* 1712:
223–45.

Roe, Shirley A. 1981. *Matter, Life, and Generation. 18th-Century*
*Embryology and the Haller-Wolff Debate.* Cambridge: Cambridge
University Press.

Roll-Hansen, Nils. 1978. "Drosophila Genetics: A Reductionist
Research Program." *Journal of the History of Biology* 11: 159–210.

Roux, Wilhelm. 1881. *Der Kampf der Theile im Organismus. Ein*
*Beitrag zur vervollständigung der mechanischen Zweckmässigkeits-*
*lehre.* Leipzig: W. Englemann.

Roux, Wilhelm. 1888. "Beiträge zur Entwickelungsmechanik
des Embryo. Über die künstliche Hervorbringung halber Embryonen durch Zerstörung einer der beiden ersten
Furchungskugeln, sowie über die Nachentwickelung (Postgeneration) der fehlenden Körperhälfte." *Virchows Archiv für*

*Pathologische Anatomie und Physiologie und für Klinische Medizin*
114: 113–53. Translated as "Contributions to the Development
of the Embryo. On the Artificial Production of One of the First
Two Blastomeres, and the Later Development (Postgeneration)
of the Missing Half of the Body." In *Foundations of Experimental
Embryology,* edited by Benjamin H. Willier and Jane M. Oppenheimer, 2–37. New York: Hafner Press, 1964.

Sánchez Alvarado, Alejandro. 2018. "To Solve Old Problems, Study
New Research Organisms." *Developmental Biology* 433: 111–14.

Skinner, Dorothy M., and John S. Cook. 1991. "New Limbs for Old:
Some Highlights in the History of Regeneration in Crustacea."
In *A History of Regeneration Research: Milestones in the Evolution
of a Science,* edited by Charles E. Dinsmore, 25–45. Cambridge:
Cambridge University Press.

Smelick, Chris, and Shawn Ahmed. 2005. "Achieving Immortality
in the C. elegans Germline." *Ageing Research Reviews* 4: 67–82.

Spallanzani, Lazzaro. 1768. https://archive.org/stream/b30356167
?ref=ol#mode/2up "Prodromo di un opera da imprimersi sopra
la rirproduzioni animali." Modena: Giovanni Montanari. Translated by Matthew Maty. 1769. "An Essay on Animal Reproductions." London: T. Becket and P. A. de Hondt.

Sunderland, Mary E. 2010. "Regeneration: Thomas Hunt Morgan's
Window into Development." *Journal of the History of Biology* 43:
325–61.

Sturtevant, Alfred H. 1959. "Thomas Hunt Morgan." *Biographical
Memoirs of the National Academy of Sciences* 33: 283–325.

Tansley, Arthur. 1935. "The Use and Abuse of Vegetation Concepts
and Terms." *Ecology* 156: 284–307.

Terrall, Mary. 2014. *Catching Nature in the Act. Réaumur and the Practice of Natural History in the Eighteenth Century*. Chicago: University of Chicago Press.

Trembley, Abraham. 1744. *Mémoirs pour servir à l'histoire d'un genre de polypes d'eau douce, à bras en forme de cornes*. Leiden: Verbeck. Translated by Howard M. Lenhoff and Sylvia G. Lenhoff. 1986. *Hydra and the Birth of Experimental Biology, 1744: Abraham Trembley's Memoirs Concerning the Natural History of a Type of Freshwater Polyp with Arms Shaped Like Horns*. Pacific Grove, CA: Boxwood Press.

Trewavas, Anthony. 2006. "A Brief History of Systems Biology. 'Every Object that Biology Studies Is a System of Systems.' Francois Jacob (1974)." *Plant Cell* 18: 2420–30.

Valli, Hanna, Bart T. Phillips, Gunapala Shetty, James A. Byrne, Amander T. Clark, Marvin L. Meistrich, and Kyle E. Orwig. 2014. "Germline Stem Cells: Toward the Regeneration of Spermatogenesis." *Fertility and Sterility* 101 (no. 1): 3–13.

Van Dyne, George M. 1966. "Ecosystems, Systems Ecology, and Systems Ecologists." Report from Oak Ridge National Laboratory.

Vicedo, Marga. 1990. "T. H. Morgan: Neither an Epistemological Empiricist nor a 'Methodological Empiricist.'" *Biology and Philosophy* 5: 293–311.

Wade, Nicholas. 1998. "Scientists Cultivate Cells at Root of Human Life." *The New York Times*. November 6, 1998: https://www.nytimes.com/1998/11/06/us/scientists-cultivate-cells-at-root-of-human-life.html.

Wang, Yuying, Ricardo M. Zayas, Tingxia Guo, and Phillip A.

Newmark. 2007. "Nanos Function Is Essential for Development and Regeneration of Planarian Germ Cells." *Proceedings of the National Academy of Sciences* 104 (no. 14): 5901–6.

Wanjek, Christopher. 2016. "Systems Biology as Defined by NIH. An Intellectual Resource for Integrative Biology." https://irp.nih.gov/catalyst/v19i6/systems-biology-as-defined-by-nih (accessed 9 August 2020).

Wheeler, William Morton. 1895. "Translation of Wilhelm Roux's 'The Problems, Methods and Scope of Developmental Mechanics.'" *Biological Lectures of the Marine Biological Laboratory, Woods Hole,* 149–90.

Wilson, Edmund Beecher. 1896. *The Cell in Development and Inheritance.* New York: Macmillan.

Wolpert, Lewis. 1991. "Morgan's Ambivalence: Gradients and Regeneration." In *A History of Regeneration Research: Milestones in the Evolution of a Science,* edited by Charles E. Dinsmore, 201–17. Cambridge: Cambridge University Press.

Yoshida, Keita, Akiko Hozumi, Nicholas Treen, Tetsushi Sakuma, Takashi Yamamoto, Maki Shirae-Kurabayashi, and Yasunori Sasakura. 2017. "Germ Cell Regeneration-mediated, Enhanced Mutagenesis in the Ascidian Ciona intestinalis Reveals Flexible Germ Cell Formation from Different Somatic Cells." *Developmental Biology* 423 (no. 2): 111–25.

Zernicka-Goetz, Magdalena, and Roger Highfield. 2020. *The Dance of Life. The New Science of How a Single Cell Becomes a Human Being.* New York: Basic Books.

# Index